Realizing the Witch

 **forms of living**

Stefanos Geroulanos and Todd Meyers, *series editors*

Realizing the Witch

Science, Cinema, and the Mastery of the Invisible

Richard Baxstrom and Todd Meyers

FORDHAM UNIVERSITY PRESS

NEW YORK 2016

Frontispiece: Svensk Filmindustri poster for *Häxan* (1922)

Copyright © 2016 Fordham University Press

Visit us online at www.fordhampress.com.

Library of Congress Cataloging-in-Publication Data

Baxstrom, Richard.
   Realizing the witch : science, cinema, and the mastery of the
invisible / Richard Baxstrom, Todd Meyers.
        pages cm. — (Forms of living)
   Includes bibliographical references and index.
   Includes filmography.
   ISBN 978-0-8232-6824-5 (hardback) —
ISBN 978-0-8232-6825-2 (paper)
   1. Häxan (Motion picture)   2. Witchcraft—Europe—
History.   3. Witches—Europe. 4. Christensen, Benjamin,
1879–1959.   I. Meyers, Todd.   II. Title.
   BF1584.E85B39 2015
   133.4'309—dc23

                                                      2015008857

Printed and bound in Great Britain by
Marston Book Services Ltd, Oxfordshire

18 17 16   5 4 3 2 1

First edition

for Julia
for Fanny

CONTENTS

x   *Contents*

What Is *Häxan*?

> Witches always claim that they do not believe in spells, object to the discourse of witchcraft, and appeal to the language of positivism.
>
> —JEANNE FAVRET-SAADA, *Deadly Words* (1980)

> It is not living life for an art form to serve merely as an echo of something else.
>
> —BENJAMIN CHRISTENSEN, "The Future of Film" (1921)

> To think is always to follow the witch's flight.
>
> —GILLES DELEUZE AND FÉLIX GUATTARI, *What Is Philosophy?* (1991)

The Wild Ride. The Sabbat. Child sacrifice. Diseases, ruin, and torture. The old hag. The kleptomaniac. The modern hysteric. Benjamin Christensen took the threads of phantasm and wove them into a film thesis that would not talk about witches, but would give the witch life. *Häxan* is a document, an amplified account of the witch insistent on its historical and anthropological qualities, presented through excesses so great that they toyed with his audience's skepticism as much as their sensitivity. Christensen created an artistic work filled with irrationalities that not only made the witch plausible, but real.

By the time Benjamin Christensen (1879–1959) began filming *Häxan* in 1921, he had already spent nearly three years conducting research for his film and securing a studio in Copenhagen to accommodate his costuming and elaborate set designs. *Häxan* was not the Danish actor/director's first foray into filmmaking, but it would be his most ambitious. The silent film was the most expensive ever produced in Scandinavia.[1] The Swedish film production company, Svensk Filmindustri, provided Christensen with funding (in addition to buying back the director's own studio facility from creditors) in 1919 to make what Christensen called "a cultural–historical film in seven acts." With Swedish funding came a Swedish title for the film (the Danish word for "witch" is *heksen*). The film was shot in Copenhagen in 1921–22, and premiered in Stockholm in September 1922. Despite laborious planning, seemingly endless trouble with censors, and the unprecedented scale of his project, *Häxan* was only the first of a trilogy imagined (but never realized) by Christensen—the other films in his unfinished series were tentatively titled *Helgeninde* (*Saints* [feminine]) and *Ånder* (*Spirits*).

Before filming, Christensen obsessively gathered historical and contemporary sources for his "lecture in moving pictures."[2] *His* "witch" would be a visual account with direct reference to original writings, art, and literature on witchcraft and witch trials from the Middle Ages through the Reformation and beyond—materials which he considered alongside and through his reading of the modern sciences of neurology, psychiatry, anthropology, and psychology. Taking up an argument made by a long line of eminent scientists—a line dominated by the preeminent psychiatrist and neurologist Jean-Martin Charcot—Christensen intended to make a case for witchcraft as misidentified nervous disease and to highlight the incompatibility of superstition and religious fanaticism with modernity and science. What resulted was something very different; it is this productive gap between intent and outcome that forms the core of what follows in this book.

Perhaps with the exception of a select group of Scandinavian film experts and enthusiasts, Benjamin Christensen is regarded—when regarded at all—as a weirdly interloping figure, much in the way his masterpiece is seen as wild and unmoored from other parts of cinema. Born the last of twelve children to a bourgeois family in Viborg, Denmark, Christensen had been active on the Danish arts and theater scene since 1902, after having trained first in medicine and then as an opera singer. Although highly praised for his singing ability, Christensen developed a nervous illness that resulted in the loss of his voice, a debilitation that led to a brief career in commercial sidelines (most notably the sale and distribution of champagne for a French importer), but this did not end his artistic career. Breaking in as a film actor in 1911, Christensen quickly moved to the director's chair; his first film, *The Mysterious X*, also known as *Sealed Orders* (*Det hemmelighedsfulde X*), was released in 1914. *The Mysterious X* foreshadowed his work in *Häxan* and was highly praised for its innovative command of cinematic technique; it was also a worry for its producers, as the film ended up costing four times its original budget. *The Mysterious X* was followed in 1916 by *Blind Justice* (*Hævnens nat*), which was celebrated for Christensen's intense performance in the tragic lead role of a man wrongfully jailed and separated from his child.

Christensen demonstrated an impressive command of filmmaking well in advance of his contemporaries. These early successes established for Christensen a reputation as an ambitious, innovative, and commercially reliable director within Scandinavian film circles. Now a proven auteur, Christensen signed with Svenska Biografteatern (renamed Svensk Filmindustri shortly after) in February 1919 and received full artistic control over his projects.[3] Rewarded for his previous efforts with unusually generous support, Christensen found the means to pursue the pioneering, bizarre, and lavish project that is the subject of this book.

Christensen's post-*Häxan* struggles are hardly surprising in retrospect. After previewing an uncensored print of *Häxan* in 1924, Metro-Goldwyn-Mayer studio head Louis B. Mayer asked, "Is that man crazy or a genius?"[4] Banking on "genius," Mayer offered Christensen the chance to make films in Hollywood, which Christensen accepted in 1926. As was often the case for European directors who moved to the major American studios in the 1920s, Christensen found that both his creative freedom and, crucially, his monetary resources were severely limited within the Hollywood studio system. After working on three lackluster projects for Metro-Goldwyn-Mayer,[5] Christensen moved to First National Pictures in 1928, directing a number of largely forgettable films, including *The Hawk's Nest* (1928), *The Haunted House* (1928), *House of Horror* (1929), and with minor praise, *Seven Footprints to Satan* (1929). While at times these films display some of the spark of Christensen's previous work, they are hardly remarkable and are today remembered only because the director of *Häxan* stood at their helm.

Christensen, in uncharacteristically quiet fashion, left Hollywood behind in 1935 and returned to Denmark. After a period of reassessment,[6] he entered filmmaking again, this time for Nordisk, producing four sound features between 1939 and 1942. While several of the "social debate" films Christensen directed at this time were critical and commercial successes,[7] the failure of his last project, the spy thriller *The Lady with the Light Gloves* (*Damen med de lyse Handsker*), brought his filmmaking career to an unceremonious close. Not even the fêted rerelease of *Häxan* to theaters in Denmark in 1941 could offset this final, humiliating disaster. After losing studio backing for an adaptation of Tove Ditlevsen's novel *A Child Was Hurt (Man gjorde et Barn fortræd)*, Christensen "retired" to manage the Rio Bio cinema in Copenhagen until his death on April 1, 1959.

Häxan stands as the filmmaker's lasting contribution to the history of cinema. And yet, it is fair to ask what precisely this masterwork *is*, because of its troubled and often clandestine status in the years since its initial release, and also because it resists easy characterization. It is typically associated today with the horror genre and simultaneously with proto-documentary or nonfiction film, but neither designation fully captures the film's unique character. Its reemergence in the middle of the twentieth century as a "cult classic" has only worked to further obscure its place within the history of cinema. *Häxan* is, perhaps more than any other work in the medium, a singular film, but for reasons that require a detailed look at Christensen's project that goes beyond the narrow lens of its seemingly strange subject matter. In the following pages we attempt *to think alongside* Christensen, through his vast library of source materials and within the structural logic of his film thesis, to better appreciate Christensen's project. And with this appreciation come questions—questions about the relationship between the image and the word, questions about the kind of new and unexpected visual historiography the film seems to produce, and finally, questions about what types of conclusions we are meant to glean from Christensen's cinematic vision.

Christensen's vision for *Häxan* was quite radical. Eschewing typical notions of "drama" and "plot," the director described his project as follows: "My film has no continuous story, no 'plot'—it could perhaps best be classified as a cultural history lecture in moving pictures. The goal has not only been to describe the witch trials simply as external events but through cultural history to throw light on the psychological causes of these witch trials by demonstrating their connections with certain abnormalities of the human psyche, abnormalities which have existed throughout history and still exist in our midst."[8] In unambiguous terms, Christensen expresses his thesis and his method in this short statement prior to the release of the film. Even if this were "all" the film did, it would still represent a significant work within the history of early nonfiction filmmaking. Yet there is so much more in *Häxan* as Christensen struggles to realize his thesis. As an enthusiastic scholar and an unusually innovative artist, he puts cinema to work not only in explicating the witch and the hysteric, but along in bringing their shared power to life.

Häxan exists as one of the most innovative films to emerge during the silent era. The film also affords a fascinating view into wider debates in the 1920s regarding the use of film in medical and scientific research, the evolving study of religion from historical and anthropological perspectives, and the complex relations between popular culture, artistic expression, and scientific ideas. *Häxan* therefore bears a unique relation to all these areas and yet is not reducible to any one of them. Christensen spent years gathering classical and anthropological sources on sorcery and religion to form the narrative background of *Häxan*. Our study of *Häxan* simultaneously offers an analysis of the scope and influence of Christensen's remarkable work and an examination of the sources that made *Häxan* a "living" cinematic tableau.

Future Forms

> Like every other artist, film artists must display in the future their own personality in their works.
>
> Benjamin Christensen, "The Future of Film" (1921)

The formal characteristics of cinema as a medium of expression provide Christensen the means to ultimately *exceed* the power of his historical source material—to breathe life into his subjects. We argue that this excess makes *Häxan*'s status complicated. While cinema has its own limitations in conveying depth and analysis in relation to reasoned arguments, it is not limited by what James Siegel has called "social[ly] constrained thinking" or a discursive mode of analysis that is, by definition, incapable of comprehending certain situations that lie outside the circle of reason. As Siegel pointedly argues, witchcraft is precisely one of the domains where scientific reasoning finds its limit; this limit is not automatic in regard to cinema, however, and Christensen exploits this fact to the fullest in *Häxan*.[9]

Most understandings of cinema would attribute this tension between reason and its absence to film's ability *to bring fantasy to life*. We would certainly not deny this claim, but it is important to note that Christensen does *not* conceive *Häxan* as a fantasy film. The director begins from the position that he will bring *errors of belief* in the fifteenth and sixteenth centuries into our view in order to show just who witches (and later hysterics) really were. Christensen's strategy is therefore more blunt than those deployed, for instance, in the human sciences, in that it *requires* the violent, erotic nature of this "error" be *emphasized* rather than *suppressed* under the signature of a theory of society as a knowable, structural entity. Jeanne Favret-Saada states it best:

> To say that one is studying beliefs about witchcraft is automatically to deny them any truth; it is just a belief, it is not true. So folklorists never ask of country people: "what are they trying to express by means of a witchcraft crisis?," but only "what are they hiding from us?" They are led on by the idea of some healer's "secret," some local trick, and describing it is enough to gratify academic curiosity. So witchcraft is no more than a body of empty recipes (boil an ox heart, prick it with a thousand pins, etc.)? Grant that sort of thing supernatural power? How gullible can you be?[10]

Christensen's thesis echoes Favret-Saada's words loudly; he is concerned (gripped) with *abnormalities*, *events*, and *causes*. The director claims, in other words, to provide a *diagnosis*, which is precisely the case when measured against the film that resulted. Yet in the details of his chosen form of expression, and through the marked intensity of his personal engagement, Benjamin Christensen does not so much unmask the witch or the hysteric as he brings these figures and the power that animates them to life on the screen. Favret-Saada's sarcasm highlights the standard objection that bringing a figure to life in this way would, by definition, violate standards of objectivity. In her ethnography of modern witchcraft in the France, the people of Favret-Saada's village insisted on the ignorance of witchcraft's power, and yet that power dictated their action and movement—to be "caught" (*pris*) or bewitched was an awareness and a risk, and as a social scientist deemed "expert" on witchcraft, Favret-Saada was an un-witcher at once caught and catching (*contagieux*) (concepts we explore in detail in our chapter on the viral character of the witch).[11] Christensen's answer, like Favret-Saada's, is that witchcraft is difficult to study because of the inaccessible materiality of the witch to the anthropologist and the believer alike.

The observable epiphenomenon signaling the presence of a witch does not yield any objective, total proof when segmented and scrutinized at an analytic distance. And yet, in the face of the failure by the human scientist to objectively signify her "true" meaning, the witch remains.

In short, bringing the witch to life "objectively" is a contradiction in terms or, to put it another way, *nonsense*. Christensen therefore only really succeeds in his aspiration by being *caught* by the power of the witch—caught up in the contest of objectivity, reason, and its necessary and instrumental negation. As we will show in detail in the introductory section to Part I of the book, there is a lengthy tradition in anthropology claiming (even in the same breath as disavowing it) that what lies at the heart of witchcraft's operations, its resistance, and ultimately the discourses of its depiction and explanation, is precisely this unattributable power.[12] The fact that cinema uniquely brings such attributes (among others) to life has led us to pursue an analysis of Christensen's film as both an *anthropological* and *cinematic* object.[13] *Häxan* reveals the logical conundrum Favret-Saada defines. Any claim to knowing the witch implies being caught by her; any claim to objectivity in the face of the witch requires her disavowal as a "real" entity. This operational paradox drives *Häxan*—and as we shall suggest, it drove efforts to locate and combat witches, and it served to shape similar aporias between sense and distance imbricated in the invention of modern hysteria in the nineteenth century and the "discovery" of the "native's point of view" by anthropologists in the early twentieth century. Taken in this context, *Häxan* is perfectly empirical and is able to express the multiplicitous character of the witch, the hysteric, and the institutions that sought to formulate knowledge regarding the real character of such entities.

Häxan: *A Film Thesis*

The idea of *Häxan* is relatively straightforward: in light of innovations in psychoanalysis and the biological sciences, Benjamin Christensen advances the thesis that the appearance of witchcraft in Europe during the late medieval and early modern periods was actually due to unrecognized manifestations of clinical hysteria and psychosis. Lacking the scientific knowledge and insight of the present age, the spectacular symptoms of hysteria (most often identified in women) were misattributed to the power of Satan and the condition of being in league with him. Deftly weaving contemporary scientific analysis and powerfully staged historical reenactments of satanic initiation, possession, and persecution, *Häxan* creatively blends spectacle and argument to make a deeply humanistic call to reevaluate both the understandings of witchcraft in European history and the contemporary treatment of hysterics and the psychologically stricken. In doing so Christensen takes on an anthropological disposition, offering *Häxan* as an expression of his own creative trials and as an empirical visual thesis *to be tested* in the world.

While we believe the above synopsis to be accurate, it only begins to characterize the complexity, innovation, excessiveness, and influence of *Häxan*. As concerned as the film is with expressing a particular idea regarding the relation between witchcraft and hysteria, it differs from many of the documentaries that come immediately after it, largely due to Christensen's explicit understanding that any idea communicated in a film must be expressed *cinematographically*. Quite unlike the sober documentary ideal that John Grierson formulated at the end of the 1920s and elaborated through the next decade via the influential British social documentary movement, *Häxan* does not conflate expressing "the real" cinematically with simple "communication." Displaying an affinity with scientists of the mid-nineteenth century (in particular, Jean-Martin Charcot, who himself claimed only "to record images"),[14] Christensen very clearly aspires to "make nature speak." What distinguishes *Häxan* is the fact that its creator had no expectation whatsoever that the real will simply "speak for itself." This distinction marks *Häxan*'s unreservedly singular approach toward conveying a particular truth about the world. Christensen's explicit refusal to either privilege a sober, generalized, and abstracted form of "truthful" visual presentation in *Häxan* or to divest the film of its serious intent despite the excessiveness of its reenactments was one of the primary sources of controversy at the time—and has continued to be an issue for the film in the decades since. Drawing freely and openly from a variety of representational strategies (scientific, historical, avant-garde, literary), *Häxan*'s status as a truthful representation was/is therefore entirely unclear. In a fashion more extreme than other ambiguously "nonfiction" films released at roughly the same time (including *The Battle of the Somme* and *Nanook of the North*), *Häxan* explicitly demonstrated the highly relational and crosscutting influences that characterized "actual" cinematic documentation and representation of the world. As such, we argue that one lasting effect of *Häxan*'s reception in the 1920s was to energize a negative, conceptually dogmatic discourse that formed and hardened cinematic taxonomies, particularly the now-"commonsense" division between "documentary" (nonfiction) and "feature" (fictional) films. Christensen's own intense fascination—his subjectively fraught relationship to the witch he was so keen to objectify—was unacceptable to nonfiction filmmakers at the time. From our vantage point today, these ambiguous, disturbing, *nonsensical* elements are some of the strongest reasons for revisiting the film and reconsidering its place within the history of nonfiction and documentary cinema.

We conceived the following pages at the interstice of cinema studies, film theory, anthropology, intellectual history, and science studies. While working in this "between," we nevertheless treat Christensen's film as *a film*, purposeful in its artistic crafting, simultaneously image *and* object.[15] Our approach is to think alongside Christensen as his film unfolds, scene by scene. Therefore, in the book we follow the narrative and structure of the film closely, dividing our text into two parts that retain and closely track the seven "cultural–historical" chapters (and thus the general structure) of the film itself. As the film has, by design, no overarching plot or main characters in the traditional sense, we feel this close, formalist strategy permits a full reading of the work. And as the film is unmistakably "biographical," that is, impossible to wrench from the events Christensen's life, these details surrounding its production are discussed. We should be clear—others have studied *Häxan* in a variety of ways (see specifically the writings of John Ernst, Jytte Jensen, Casper Tybjerg, Arne Lunde, and Jack Stevenson), however none have fully explored the film through the theoretical framework that Christensen attempts to construct regarding the witch's material, invisible, mobile force.[16] We insist that in order to appreciate Christensen's vision, it is not enough to think *about* the film, one must think *with* it—to allow oneself to be ensnared by the witch, as Christensen indeed was.

We begin Part I by introducing the historical and epistemological contexts within which Benjamin Christensen's *Häxan* emerged, specifically engaging the source material that guided the film's treatment of irrationality and "nonsense" to form a conceptual framework as a necessary starting point for our analysis. Turning to the film itself, Chapters 1 and 2 are concerned

Hans Baldung Grien, *Witches' Sabbath* (1510). Courtesy of British Museum, London.

with *evidence*—words and things, the theological debates that establish the groundwork for Christensen's thesis, and the forms of evidence found in the writings on witchcraft and sorcery at the time used by inquisitors to identify or "name the witch." In these chapters we explore the cinematic forms evidence takes—the still image, the vignette, the reenactment, the facial close-up—which Christensen molds as visual strategies, making them tactile, twisting and bending the natural order at the level of the profane in order to show the ways in which the witch is realized.

In Chapter 3, we explore the viral character of the witch. In particular, we highlight how witchcraft allegations were given signification in early modern Europe and how the increasingly formalized criteria as the basis of witchcraft evidence tended to multiply and spread, rather than reduce, the number of individuals who were suspected of being in league with the Devil. The epidemic atmosphere of accusation required and was enabled by forms of experimental practice, and it is through the emphasis on *experimentation* and *evidence making* that *Häxan* finds common ground with other nonfiction films and the production of scientific images during the nineteenth and twentieth centuries. In this chapter Christensen demonstrates how witch trials were structured as an experimental process whereby inquisitors and laypeople would labor to establish *proof* of something they could not yet see but knew to be present. Here, *Häxan* employs the well-established trope of sickness and diagnosis to demonstrate how *maleficium* was detected and confronted.

In Chapter 4, we highlight the traces of Christensen's own demonological thinking. It is in this chapter of the film that Christensen begins to draw links between diagnostic strategies for identifying nervous disease and the forms of interrogation used by magistrates outlined in guides such as the infamous *Malleus Maleficarum* (1487). Christensen takes inventory of the elements that together created the witch stereotype in the early sixteenth century, including the Wild Ride, the pact with the Devil solemnized through sexual intercourse, cannibalism, and the cauldron.

We begin Part II with an introductory section titled "1922," where we explore activities in other parts of the arts and human sciences around the time of *Häxan*'s creation that shed light on Christensen's deep commitment to the power of evidentiary thinking. In Chapter 5 we turn to the concepts and uses of sex, touch, and materiality in order to examine the personalized, elaborated visualization of the witch stereotype realized by Christensen in the previous chapters of the film. In this chapter, primary accounts such as Johann Weyer's *De praestigiis daemonium* (1563) serve to reveal the complex nature of sensual explorations of the flesh through masochism and exorcism.

In Chapter 6, the appearance of possession, ecstasy, and "insanity" reveals the reach of demonic influence and introduces the important genres of transfiguration and metamorphosis. In this chapter of the film, Christensen for the first time uses testimonies to further his cinematic thesis. The chapter showcases instruments and techniques of torture in order to highlight the highly charged expression of religious ecstasy and self-mutilation as part of his visual tableau. It is here that Christensen draws from his most inexhaustible resource: the neurological writings of Jean-Martin Charcot and his followers, especially the volumes they produced in the *Bibliothèque diabolique* that explicitly dealt with the relationship between witchcraft and nervous disease.[17]

In the final moments in his film, Christensen is clearly anxious to draw the threads of his thesis together at the close. He makes the most explicit overtures to show how those who were once identified as witches are now the objects of medical and social concern in modern life. Despite his effort, what results in the conclusion is something quite different. Christensen does not dispense with the witch as an aberration from the past haunting the unfortunate and superstitious in the present, but instead shows the potency of her various forms over time. In our postscript we use the ambiguity of Christensen's suppositions to consider *Häxan*'s innovation and singularity within the silent period, the historical and social contingencies that surrounded the film, and its place within the history of cinema more generally.

We seek to understand *Häxan* on its own terms. While this sounds straightforward, it is not without some blind alleys and contradictions. Our narrative is woven in and through Christensen's historical sources and theoretical commitments, as well as into the technical realities and innovations of his filmmaking. We adopt the existing organizational structure of the film in an effort to show what drives Christensen's thesis, often in ways that perplex as much as inform viewers. While we certainly seek to clarify rather than confuse, we feel it crucial to allow the film and its thesis to stand. It should be obvious by the end that we regard *Häxan* to be vitally important as a film precisely due to Christensen's ability to bring the witch to life while retaining the mysterious core that gives her life in the first place. Like Christensen, we cannot avoid being caught a little by the witch, seeking at some level to only pass along the experience of being rapt by this work of cinematic art in a manner proper to our own positions as anthropologists and human scientists.

Albrecht Dürer, *The Four Witches* (1497). Courtesy of British Museum, London.

PART ONE

The Realization of the Witch

The Witch in the Human Sciences
and the Mastery of Nonsense

> Beloved, believe not every spirit, but try the spirits whether they are of
> God: because many false prophets are gone out into the world.
>
> —I JOHN 4:1

There is a largely unacknowledged historical tendency and predisposition
within the human sciences with roots in much older practices of defining
social facts and the discovery, interpretation, and the production of the real
itself. In plain language, it emerges from a method that allows the researcher
to sense, interpret, and eventually master forces that appear to be nonsensical and yet are held to be essential to the reality of everyday social life. While
such invisible forces have gone by many names, one can track a historical
persistence of this epistemological concern with things that cannot be seen
or logically interpreted but are nevertheless held to be present.[1]

One way of tracking this problem of the mastery of invisible forces has
been offered by the literary scholar Jonathan Strauss, specifically with regard to the notion of the irrational as a privileged space in medical discourses
in nineteenth-century Paris. Strauss argues that the role of irrationality and
"nonsense" was a "legitimizing force" for medicine in that "the very incom-
prehensibility of the mad created a mysterious and extra-social language that

the rising medical profession could adapt to its own purposes."[2] This kind of mastery is of course no news to anthropologists, who have claimed a similarly privileged space in the late nineteenth and early twentieth centuries through their understanding of the "nonsense" of "the native." The empirical mastery of domains consigned to the illogical realm of human social life—and in particular life in distant societies—formed the methodological basis that allowed the fieldworker "to see" unknown forces. From Malinowski forward, ethnology depended on exactly this process, as anthropologists forged a bond with the invisible and irrational as a methodological pillar. Anthropologists thus had to develop a battery of tests that could yield some felicitous information as to the "true" nature of unseen forces and their operations within empirical, real-world contexts. The heart of our argument in this book is that *Häxan*, in its curiously excessive attempt to produce a nonfiction film about the power of the witch, deploys an analogous approach and relies on very similar conceits for citing evidence of what is empirically "real" in the world.

The attempt to secure evidence of forces felt but unseen is certainly not an invention of the nineteenth-century sciences of life and man.[3] A clear conceptual link exists between the investigative techniques developed by sixteenth-century theologians and Church inquisitors in the face of what was understood to be a vast proliferation of the incredible, unbelievable power of Satan and emergent scientific fieldwork practices in anthropology and other social sciences in the late nineteenth and early twentieth centuries. While the systematic, empirical investigation of strange events, singularities, miracles, and other types of staple phenomena in preternatural philosophy predates Francis Bacon's *The Advancement of Learning* (1605), there is a method that emerges within the ensemble of human sciences proper to the science of *man* that is unable to expel these direct, necessary engagements with unseen and empirically unprovable forces.[4] Although the credible status of such phenomena as real per se has been detached from these disciplines, the status of these phenomena as dark precursors[5] driving the inquiries taken through the signatures of anthropology and science serves as the focus of our engagement here. As such, we argue that anthropology as a science is predicated on rationally mastering invisible, irrational forces. Or, perhaps more precisely, anthropology emerges as a distinct human science from the desire to credibly master *nonsense*. Well versed in anthropological literature regarding witchcraft, possession, and ritual, Benjamin Christensen, too, demonstrates the desire to bring the invisible and nonsensical into view; although Christensen's medium was cinema rather than more traditional forms of ethnological record, *Häxan* nevertheless stands as one of the most powerful, unsettling expressions of the aspiration to produce evidence of forces unseen.

Myths, Origins, and Methods

Following what George Stocking has termed the "Euhemerist Myth" of anthropology[6]—that is, a rationalizing tendency to interpret mythology as historical event—we argue that the links between Christensen's *Häxan* and Bronislaw Malinowski's fabled definition of the methodological task of the anthropologist are undeniable. In the ur-text of this myth, *Argonauts of the Western Pacific*, Malinowski confidently identifies "the final goal, of which an Ethnographer should never lose sight":

> This goal is, briefly, to grasp the native's point of view, his relation to life, to realize his vision of his world. We have to study man, and we must study what concerns him most intimately, that is, the hold which life has on him. In each culture, the values are slightly different; people aspire after different aims, follow different impulses, yearn after a different form of happiness. In each culture, we find different institutions in which man pursues his life-interest, different customs by which he satisfies his aspirations, different codes of law and morality which reward his virtues or punish his defections. To study the institutions, customs, and codes or to study the behavior and mentality without the subjective desire of feeling by what these people live, of realizing the substance of their happiness—is, in my opinion, to miss the greatest reward which we can hope to obtain from the study of man.[7]

Although subjected to rigorous critique in the decades since its original publication in 1922 (the same year *Häxan* was released), Malinowski's direct expression of the desirable method and the underlying aspiration grounding this technique has never been definitively overturned within the discipline. To this day the paragraph quoted above serves as the distillation of method and disposition alike when confronted with the deceptively difficult questions

"Who are you?" and "What do you do?" The assertion by anthropologists claiming to have assumed the "point of view" of another, not to mention the resulting ethical disequilibrium, has been rightly subjected to a series of stringent critiques over the years. But the idea that we should fully dispense with Malinowski's epistemological aspiration and regard interlocutor others as "Other" remains unthinkable within the discipline as well.[8] This inconsistency has generally been resolved by one of two potential displacements: the first proposes that we detect the underlying structures framing "points of view," while the second aims to appreciate the meaning of social facts as a substitution for Malinowski's blunt demand to assume the simultaneous position of the "social" scientist and the object of this science.

What grounds Malinowski's claim that the fieldworker must achieve the cultivated, sensed point of view of another is a privileged relation to the unknown. This privileged relation must emerge through experimentation and through the ability to, in some fashion, test what is asserted to be real; in the anthropology of Malinowski's vision this test is a series of subjective trials[9] subsumed within the rubric of "fieldwork." In this way, a discipline such as anthropology can legitimately claim kinship not only with other human sciences but also with the "hard" sciences. The tie between mastering what Strauss has termed "nonsense" and asserting scientific authority has strong links to transformations that occurred in the course of the "witch craze" in Europe, specifically regarding the terms of evidence within the overlapping institutional domains of science and law, both dominated by theology, to which we will return in the following pages. Certainly institutions charged with the task of discerning truth from falsehood have shifted dramatically over the centuries, yet the murmurs of this original theology remain audible in Malinowski and Christensen, even today. *Häxan* exists as a visual amplifier of these persistent murmurs.

Malinowski's method requires certain presuppositions in order to be effective. First, it presumes that the experiential disposition of the analyst is a legitimate and effective way by which one can begin to form an understanding of a phenomenon otherwise held to be imaginary, fictional, or simply untrue. Second, it turns on the principle that witnessing and testimony can concretely serve as evidence as to the reality of something otherwise beyond the direct experience of the researcher. In seeking to bring the invisible and nonsensical into the realm of ethnographic fact, Malinowski explicitly recognizes the representational nature of this truth; only the testimony of the expert makes belief in such phenomena as real (in any sense) possible. Dan Sperber has pointed out that most religious beliefs follow the same representational logic.[10] Since Luther's radical assertion that faith can only be a commitment to the representation of a truth, the explicit nature of this relationship has been a contentious element in Western Christianity's own efforts to discern truth and the nature of the world. Malinowski has thus only updated and secularized a much older epistemology dating back to precisely the period *Häxan* depicts. In the words of Joseph Leo Koerner, "Lutheranism is the original anthropology of 'apparently irrational beliefs.'"[11] As we shall see in *Häxan* (a quite "Protestant" work in many ways), this overriding "conviction in the utterly invisible" is not solely the concern of either theologians or scientists and hardly limited to the time of the Reformation and the subsequent witch craze.

Realizing the Witch

In the closing decades of the fifteenth century, it was clear to ecclesiastical and secular authorities in Europe that they were witnessing a crisis in the form of a proliferation of witches.[12] The growing number of beings intent on the destruction of Christendom mirrored the growing power of Satan on earth, and for many, indicated an impending apocalypse. In more immediate terms for theologians, the seemingly viral proliferation of demonic power beyond the grasp of human experience, intuition, or thought required a radical change in the manner by which authorities could investigate and evaluate situations that involved invisible, supernatural powers.

First appearing in 1487, in an atmosphere of fear and grave doubt, the notorious demonological text the *Malleus Maleficarum* (*Der Hexenhammer* or *The Hammer of Witches*) established a logical if not disputed relation between investigative procedures, the constitution of evidence, and the assertion of fact during the period.[13] Proceeding in a manner explicitly contrary to previous scholastic methods of ascertaining the nature of the real, the assertion of expertise in the *Malleus* by authors Henry Institoris (Heinrich Kramer) and Jacob Sprenger, while quite radical for its time, echoes to a startling degree much later statements to the same effect, including

Malinowski's own assertions discussed earlier. The claim to expertise in the *Malleus* is phrased as follows:

> We are now laboring at subject matter involving morality, and for this reason it is not necessary to dwell on various arguments and explanations everywhere, since the topics that will follow in the chapters have been sufficiently discussed in the preceding questions. Therefore, we beseech the reader in the name of God not to ask for an explanation of all matters, when suitable likelihood is sufficient if facts that are generally agreed to be true either on the basis of one's own experience from seeing or hearing or on the basis of the accounts given by trustworthy witnesses are adduced.[14]

Institoris and Sprenger were actively responding to concrete fears of Europeans at the time. Their bold assertion of expertise in matters real but (often) invisible shares much with Luther's reply to the question of how might we see God: "Just as our Lord God is the thesis of the Decalogue, so the devil is its antithesis."[15] Nothing troubled the soul of the late-fifteenth- and then sixteenth-century European as much as God's apparent absence in times of great change and strife. Forcing Satan and his followers from the shadows through an interpretive expertise over the concrete, secondary manifestations of his reality was often reassuring, *relief* for the pious believer on the brink of doubt. Heretics such as the Brethren of the Free Spirit, Waldensians, and Cathars managed God's absence without positing the embrace of life that the Devil urges in binary opposition to that of the Good, albeit infused with the perilous dogmatism eschatology always brings.[16] Most had no luxury to imagine such a world.

As Satan's power appeared to grow (at least in the treatises of theologians) the problem of the Devil interfering with the most intimate communications with the Divine became acute.[17] How does one know who *really* hears the prayers and entreaties of the faithful? Moreover, given the Devil's deceit and omnipresence, how does one *really* know who is speaking when prayer is returned? The paradoxical comfort the inquisitor offered was rooted in questions of theodicy in a world where the trappings of belief are everywhere but where there is no incontrovertibly visible evidence of God's acknowledgment or answer to the prayers of the faithful. Thomas Aquinas had earlier raised this thorny problem of doubt: "It seems that there is no God. For if, of two mutually exclusive things, one were to exist without limit, the other would cease to exist. But by the word 'God' is implied some limit-less good. If then God existed, nobody would ever encounter evil. But evil is encountered in the world. God therefore does not exist."[18]

Aquinas refutes his own speculative preposition through his famous five proofs of God's existence;[19] demonologists of the fifteenth and sixteenth centuries were not so sure. For demonologists such as Institoris, Sprenger, or Johannes Nider, a third figure beyond that of "God" or "man" was required; this figure in concrete terms was the witch.[20] Thus, the absent term in this understanding is shifted from God (although most could not claim to have directly seen God) to the witch, the chasm between God and man now it-self functioning as a kind of proof, a reassurance that the evil of the world can be explained through the various iterations of Satan's power.[21] The Devil therefore serves to prove God's existence, the polarity reversed *toward* God's permission for demons to cause evil in the world and *away* from the nag-ging, perceived void where God is expected to be. As demonologists would persistently claim in the sixteenth century, *God must exist because Satan is right in front of me!*[22]

If human beings were slow to recognize divinity compared to malicious beings such as demons (after all, it was *demons* who first recognized the di-vinity of Christ, long before his disciples came around),[23] then how could one confidently recognize the presence of Satan? By definition inquisitors would have taken the reality of witches and Satan for granted, yet the scope of demonic power authorizing these beings concrete reality in the world would have nevertheless struck inquisitors as unbelievable.[24] Hearing the name of the witch in an accusation or a confession, bolstered by the details of truly sacrilegious and inhuman deeds, would still have been a shock to them and was very much subject to verification. Put differently: with the interweaving of learned demonology into the fabric of a dominant theology that ratified the sovereignty of God primarily through the worldly evidence of Satan's forceful opposition to that divine power, inquisitors believed that what was reported to them was possible; but it would be a gross misrepre-sentation to argue that inquisitors would not have then sought to empiri-cally verify such claims. Indeed, even within this style of reasoning, it was possible that individual accusations could be found to be spurious or false. The invisibility of the spiritual world was expressed as an essential given, but demonologists and inquisitors at the time still desired *proof.* As doubt arose everywhere around them, the viral proliferation of the witch came to provide that proof.

As numerous scholars of the witch trials have noted, the strategy of leading the accused in her testimony was common during interrogations.[25] In an effort to prove a particular instance of witchcraft had occurred, inquisitors often had to lead, goad, and viciously repeat the torture of "the witch" until a narrative was produced that at least partially satisfied the demands of evidence.[26] For the inquisitor or witch hunter, it was never enough to simply "believe." Rather, the interrogation under torture represented an experimental form of knowing in crisis.[27]

It would be absurd to argue that this style of interrogation was later simply reproduced in the more modern contexts of the human sciences or in early ethnographic studies such as Malinowski's pioneering work in the Trobriand Islands. Yet the truth value of a nonsensical confession made sensible has a strong connection to a series of truths regarding human belief, action, and social practices across a much longer historical arc than generally acknowledged. This link is perhaps even clearer if we shift our attention from the pragmatic humanism of Malinowski's approach to the ethnographic style of early French ethnographers such as Marcel Griaule. While rejecting the stark ontological difference asserted by Lucien Lévy-Bruhl between the nonsensical world of "primitives" and the science of the West, Griaule's own approach to ethnographic research developed in the 1930s betrayed an aggressive belief that "natives" could not (or simply would not) ever be able to produce a "proper" explanation of the forces around them or their own beliefs and motivations in relation to these forces. They would lie, conceal, protect—and so wresting their knowledge from them, learning truth from lie, was essential to representing their reality in order to interpret it in its true picture:

> The role of the sleuth of social facts is often comparable to that of the detective or examining magistrate. The crime is the fact, the guilty party the interlocutor, and accomplices are all the members of this society. This multiplicity of responsible parties, the extent of the areas where they act, the abundance of pieces of evidence serving to convict appear to facilitate the inquest, but in reality they guide it into labyrinths—labyrinths that are often organized. . . . Not to guide the inquest is to allow the instinctive need that the informer has to dissimulate the most delicate points. . . . The inquest must be treated like a strategic operation.[28]

Thus, while testimony was an essential tool for ethnographers of this school, the encounter between researcher and subject constituted a series of *severe tests* by which the researcher could gather the necessary empirical evidence in order to make a felicitous truth statement regarding what was "really" at play. While the nonsense to be mastered had shifted from the demonic, incredible forces at play for the inquisitor to the misguided tall tales of the native interlocutor, the logic of gathering evidence through a series of trials or tests is surprisingly durable between these investigative contexts.[29]

As authoritarian as Griaule's approach to fieldwork explicitly was, it was also consistent in its recognition of the struggle that lay at the heart of raising testimony to the status of the "really real."[30] Haunted by the possibility of deception, Griaule was more explicit in his recognition that any form of testimony (his or an interlocutor other's) requires a test or a trial in order for it to be elevated to the status of a fact. Avital Ronell captures this necessity when she writes, "A passion or experience without mastery, without subjectivity, testimony, as passion, always renders itself vulnerable to doubt."[31] Aspiring to an objective form of scientific knowledge that obviates this doubt, Griaule aggressively frames the scene of ethnographic encounter itself as a kind of antagonistic trial whereby the ghosts and gods of the natives are forced out of the shadows and made concretely apparent to the senses of the anthropologist. This approach attempts a more delicate balance than its belligerent tone would lead us to believe. Griaule himself appears to acknowledge that the testimony obtained, while able to generate some understanding, will never resolve itself in a proof in the strict sense of the term. In essence, fieldwork in this context produces knowledge of hauntings that is itself haunted. As inquisitors also tacitly acknowledged, this paradoxical haunting is what gives testimony its power of fact in the first place. If testimony were truly "certainty" or mere "information," as Jacques Derrida reminds us, testimony "would lose its function as testimony. In order to remain testimony, it must therefore allow itself to be haunted. It must allow itself to be parasitized by precisely what it excludes from its inner depths, the possibility, at least, of literature."[32]

The expertise that comes with wrangling the invisible and nonsensical very often is rendered visually (it should come as no surprise that the filmmaker Jean Rouch was one of Griaule's students). Yet strategies of visualization are hardly limited to a "French" approach in this instance, as commentators ranging from Clifford Geertz to Anna Grimshaw have noted the visual qualities of Malinowski's ethnographic writing, with Geertz going so far as to playfully term his output the imaginative result of "I-witnessing."[33] Nor are such creative test results solely a phenomenon of twentieth-century social science. The paradoxical necessity of an *expressive* element within an objective test in relation to what would otherwise be nonsense is evident in many of the examples of sixteenth-century visual culture that remain known to us today. For example, in Franz Heinemann's 1900 *Rites and Rights in the German Past* (a work that figures prominently in *Häxan*),[34] a woodcut shows an investigative technique deployed by inquisitors and witch hunters: trial by ordeal. The woodcut depicts a crowd of people surrounding a bound woman as she is nudged away from the shore. Heinemann's image is similar to others, including a detail of a bound, naked woman undergoing the trial by water drawn from Eduard Fuchs's *Illustrated Social History from the Middle Ages to the Present*.[35] The possible outcomes were few: if the woman floats she is clearly able to contravene the laws and God and nature and is therefore a witch or heretic; if she sinks, she has made no such pact with Satan and the judges proceed to thank God for her innocence (though she may have just as likely been fished out before drowning). It is important to note the role of procedural expertise that such ordeals required, as the trial by water here functions as experiment as much as a punishment, designed to reveal an otherwise invisible truth.

It is clear that concerns about what was admissible as evidence of the real motivated such "trials" and served to frame the possible interpretations of their results. Testimony, experimental results, and expert inquisitorial interpretation together came to form an early version of the *case study* that, in turn, could be synthesized as evidence in service of accounting for variation that exceeded *general laws* regarding relations and phenomenon in the world. Individual cases came to serve as an effective strategy in providing analytic purchase for phenomenon that were otherwise invisible to even the discerning eye of the expert. It is this act of taking a single, natural object (the case) and abstracting its qualities to describe phenomena that we find in the clinical work of Thomas Sydenham in 1668, something the historian Philippe Huneman further traces through the psychiatry of Philippe Pinel, which we can extend even further in the famous cases histories of Sigmund

Freud.[36] Following this thinking, close analysis of salient individual cases would make hidden tendencies visible in practical, "natural" terms, a characteristic that made the method attractive to artists and scientists alike who were seeking to move away from a reliance on metaphysics.[37] It is no accident that in the nineteenth century the clinical photography of Guillaume Duchenne de Boulogne and Jean-Martin Charcot, and the chronophotography of Eadweard Muybridge and Étienne-Jules Marey, exerted a formal, expressive influence that often exceeded the limited audience of scientific peers.[38]

Charcot and the Bibliothèque diabolique

> It is easier for superstitious men, in a superstitious age, to change all the notions that are associated with their rites, than to free their minds from their influence. Religions never truly perish, except by natural decay.
>
> W.E.H. Lecky, *History of the Rise and Influence of the Spirit of Rationalism in Europe* (1865)

> The physician seeks to fill what he knows with what he sees. He is in search of the manifestation of his nosological concepts. Mobilized by attention, he considers the deployment of a knowledge in the new and visible form of an appearing. In short, he discovers without learning.
>
> Michel de Certeau, *The Possession at Loudun* (1986)

There are witch confessions that are insane. This fact was recognized by many skeptics in the sixteenth century who, while acknowledging Satan's unquestioned power, cast doubt on the truth value of unlearned witnesses to this invisible power and the theological frameworks deployed by inquisitors validating their interpretations of how such reported acts were consistent with the authoritative discourses of the Church or (in the case of Protestants) of the gospels themselves. *Possessions* set the stage for the explicit medicalization of the mobile, invisible forces that experts had been struggling to master, explain, and take measures against—a new mode that is equally didactic and forensic. In such well-known incidents as the possessions among the Ursuline nuns of Loudun from 1634, we find an increased medicalization of the invisible that, over the course of a long transition,[39] reverberates through the medical and human sciences of the nineteenth century.

The quote from Michel de Certeau's *The Possession at Loudun* refers to the physicians called upon for aid in the wake of the Church's failure to exorcise the demons haunting the nuns at Loudun. Jean-Martin Charcot was of course well aware of the enduring relationship between religious sense and medical knowledge underscored by de Certeau's statement.[40] Thus Charcot did not so much invent as inherit a perspective on the relation between religious ecstasy, magic, witchcraft, and "nervous disease." He and his students collected and published historical accounts under the title *Bibliothèque diabolique*—a series in which many texts and treatises from the sixteenth century were reproduced, including *Soeur Jeanne des Anges, superiéure des Ursulines de Loudun: Autobiographie d'une hystérique possédée*,[41] *Science et miracle: Louise Lateau ou la stigmatisée belge*,[42] and *La possession de Jeanne Fery*,[43] all of which were accounts contemporary to Johann Weyer's *De praestigiis daemonum et incantationibus ac venificiis* (*On the Illusions of the Demons and on Spells and Poisons*, 1563), and follow the *Malleus* by nearly a century. The books in the *Bibliothèque diabolique* indexed as much as clarified the link between witchcraft and hysteria for Charcot and his followers. In *Science et miracle* (a 1875 book on witchcraft, faith healing, and demonic possession), Bourneville begins by warning his readers that the "profound time of ignorance in the Middle Ages" has been prolonged into "modern society,"[44] appealing to an appraisal of history in service of a project for scientific modernity. For Bourneville and those producing work in the *Bibliothèque diabolique*, case studies were meant to demonstrate the precariousness of misrepresentation and the consequences of ignorance.

The errors of demonologists and exorcists were rooted in what was characterized as the mistaken conceptualization of their object of investigation. Yet, accusations of error and superstition aside, the procedural elements of the investigations collected in the *Bibliothèque diabolique* bore a startling resemblance to those undertaken by Charcot and his students, particularly in their studies of hysteria. While it is certain that witch hunting and the exorcism of spirits in the sixteenth century were hardly interchangeable with clinical studies of nervous illness in the nineteenth century, the conceptual scaffolding of the emergent science that they were creating bore more than a passing resemblance methodologically to these now antiquated forms of inquiry. More than anything, the continued fascination with the secondary, visible effects of primary invisible forces demonstrated that the discernment of spirits, like the diagnosis of nervous illness, involved a long-term labor of social interpretation that required the mutation of old categories and the creation of new ones.[45] The contentious fragility of these endeavors revealed by these historical accounts served as salutary lessons for Charcot and his followers; the fact that their own conceptualization of their object largely retained its status as the insensible, invisible, outside forces that served as the focus of the work in the *Bibliothèque diabolique* was an irony that largely escaped comment.

It is impossible to overstate the influence the works produced by the students of Jean-Martin Charcot had on the trajectory of inquiry across the human sciences of the late nineteenth century. Charcot himself collected artistic and historical materials on the relation between witchcraft and hysteria, which he presented under the title *Les démoniaques dans l'art*, published by the Academy of Medicine in 1887.[46] Charcot's famous students such as Georges Gilles de la Tourette and Paul Auguste Sollier attempted "to trace the hysteric through history" with "sincerity and veracity." They concerned themselves not only with prevailing social attitudes toward "misdiagnosed" hysterics of the early modern period, but also with clinical attention to the physical manifestation of hysteria found in images and writings on the time. In Gilles de la Tourette's *Traité clinique et thérapeutique de l'hystérie d'après l'enseignement de la Salpêtrière*[47] and Sollier's *Genèse et nature de l'hystérie, recherches cliniques et expérimentales de psycho-physiologie*,[48] we find detailed indexing of symptoms such as religious fervor and stigmatization alongside psychosomatic indicators such as blue edema or swelling with local cyanosis, and "autographic skin" that would appear intensely red after touch—all physical signs of witchcraft attributed to earlier centuries.[49]

As A. R. G. Owen points out, the word "medicine" finds its etymological roots in sorcery, after Seneca's tragedy of *Medea*, whose betrayal and revenge leads to the murder of her children.[50] Even within the writings attributed to Hippocrates, "the sacred disease" (epilepsy), erroneously perceived as resulting from hostile magic, could be reconsidered in terms of individual physiological disorder.[51] Yet "hysteria," itself from the Greek for "uterus," seemed to hold a special place in the moral imaginary.[52] In the century before Charcot's famous neurology clinic at Salpêtrière, nearly ten thousand women were kept there at the second Bastille, La Force Prison. These were destitute women, the insane, "idiots," epileptics, and Parisian society's "least favored classes."[53] The special susceptibility of women to witchcraft mirrored the "feminine weakness" associated with the hysteric, exacerbated by low social status. It was nuns and devoted female members of the Church that raised special concern when "possessed" by unexplained forces of demonic or psychic origin. Like later diagnostics of hysteria, the discernment of spirits was at its root a discernment of *female bodies as such.*[54]

Ulrich Baer points out that what Charcot created was a *tableau vivant* transformed into a *tableau clinique*—a hysterical reliving of the original symptom and reframed trauma in an attempt to suspend the two temporalities (real and reimagined) in the same image.[55] This "reliving" is precisely what Freud and Breuer meant to produce through hypnosis in their studies on hysteria, to isolate the mechanisms of hysteria and the surrounding symptoms of catharsis and dementia, their most famous cases being Anna O., Frau Emmy von N., and later of course Freud's own Dora.[56] It's no wonder that one element of fascination with hysteria was its "look"—its *aesthetic* link to forms of possession. Traugott Oesterreich, who published his *Occultism and Modern Science* in 1923, traced a similar path toward an aesthetic ideal of possession beginning with the *Acta Sanctorum* in the Catholic Church.[57] In countless accounts of possession, we find descriptions of demons speaking through the mouths of girls, as well as possession manifested through "external signs" of a new physiognomy, particularly in the face. We also find identical descriptions of voice, personality, and facial change in Pierre Janet's *Névroses et idées fixes.*[58] In his studies of medical psychology, Janet describes the aneasthesias, amnesias, subconscious acts, somnambulisms, and fixed ideas all associated with possession—including a case of spontaneous abortion brought on by powerful thoughts of a previous abortion.[59] Oesterreich's observation that Catholic religious ceremonies to "treat" the possessed worked to heighten the intensity of possession, reflects the same heightening that Charcot himself considered of the subjective state of the hysteric in his clinical theater.[60]

In sum, what is critical to note in all these cases is an overriding desire to gain some empirical purchase over forces openly acknowledged to be invisible and insensible. The tension animating each of these domains lay in the conceptually arranged chasm between outer and inner states. For exorcists, building on the techniques of inquisitors and witch hunters, possession acts as the bridge across this chasm. Two centuries later, neurologists and psychologists construct the same scaffolding. Anthropologists by the time of Malinowski attempt to close this aporia by sympathetically occupying the very inner space of their interlocutors, with the witches, spirits, and demons no longer explicitly the target of the inquiry, but rather fieldworkers as truth-tellers returning from the dark corners of the real. In the midst of these efforts, *Häxan*, aspiring simultaneously to the status of science *and* of art, sought to force *everything* into plain view. As powerful and forward-looking as *Häxan* truly is, the specter of sheer nonsense as its real object remains. And yet we still today hunt ghosts and witches, fueled by a desire operationalized in a method of being close enough to something to *sense* it, because even objective scientific mastery demands a closeness to things unseen, unprovable, indeed *nonsensical*, yet *there*.

Evidence, First Movement: Words and Things

> The perfect photoplay leaves no doubts, offers no explanations, starts nothing it cannot finish.
>
> —HENRY ALBERT PHILLIPS, *The Photodrama* (1914)

> The technical structure of the archiving archive also determines the structure of the archivable content even in its very coming into existence and in its relationship to the future. The archivization produces as much as it records the event.
>
> —JACQUES DERRIDA, *Archive Fever* (1998)

In the beginning there is a word. That word is "Häxan."

Benjamin Christensen's biblical echo is intentional. From the first frame of *Häxan*, Christensen is seeking to dismantle the conventional cinematic image. This is an image of a word. In light of what is to follow, the formal conventions of the silent film by definition destabilize any easy relation to the object "Häxan"; it exists multiply. Already reaching into his source material, Christensen borrows Italian inquisitor Zacharia Visconti's categories of language to show us how the word relates to meaning, expressed in the distance between the thing and the thing signified. Visconti designated this "the language of the voice," the language proper to humans.[1] Yet, in a silent film there is no obvious voice. Certainly, a printed word occupies the domain of language for Visconti, but for this to be formally consistent the word requires the syntax that would allow the reader to insert her inner voice, the memory of a voice, in order to make this so. "Häxan" appears to lack this syntactic force at this opening instant to be properly a statement.

"Häxan"—the witch—appears to be an impossible object. In Visconti's schema, this word also appears to speak "the language of the mind." This is a language the inquisitor reserves for angels, a language resulting in non-statements. From the very beginning, there is no claim made about the witch—no question is asked. The witch is simply announced. In an instant, "Häxan"—the witch—is there and this is all.

"Häxan" is simultaneously a word, an image, and a thing. Benjamin Christensen makes every effort to craft a witch that is real to us. It is a grand ambition. Playing with the ontological fluidity of a cinematic image, the director expresses himself through an image-world that seems entirely of his own creation. "Häxan" in the opening moment of the film is a monad, containing the totality of this world in its most basic element. Not just a word, *the Word*. Visconti reserved this language, "the language of things" to God alone; yet for scientists and filmmakers, it is the language of things that holds the greatest appeal.

To the World a Witch

Christensen's first task is to open the world of the witch to the film's audience. He does this by immediately following the word with a preposition, albeit still denying us the calming language of voice that is proper to us. This preposition, denoting both agency and possession, comes in the form of a face. *His* face. The commanding, scowling face of the director stares out at the camera. Christensen's film will make full use of this art of *metoposcopy*. Dating back to Girolamo Cardano and the Renaissance, metoposcopy defined the operation of reason as the weaving together of images in the mind. In turn, the expression of reasoning was to be found on the face (a proto-cinematic theory of the relation between image and thought if ever there was one).[2] Christensen's face is one of many revealed; these faces—of the old woman, of the ecstatic nun, of the novice sorceress—will be offered as primary evidence of the power of the witch and the logic of demonological thinking. It is telling that Christensen's face is the first shown, not in order to place his seal of authorship, but as a way to assert to his audience that it is *his* argument that resides in the foreground. This is no ordinary film. It is not merely entertainment. *Häxan* is a *thesis*.

Benjamin Christensen, *Häxan*, film still (Svensk Filmindustri, 1922).

After this dramatic beginning, Christensen provides some immediate re-
prieve through a scarcely noticeable addendum to the opening title card: "A
presentation from a cultural and historical point of view in seven chapters
of moving pictures." Claiming a reassuring authority, Christensen now signals that he intends to enlighten us in the manner of a professor giving a
lecture. The technology of the motion picture is not simply a medium here;
in the service of Christensen's thesis, it is a precise, deliberate *method*.

The title cards that follow identify the director, the cinematographer, and
turn the audience's attention toward the list of sources for the film distributed as part of the original program (which has been reproduced in the back
matter of this volume). Like any respectable scholar, Christensen indexes
himself through his sources. Yet his mode of citation is unambiguously
rooted in the formal elements of *cinema* and the *image* rather than *texts*,
and is ultimately put to different uses from that of the historian or human
scientist; this difference will constitute the focus of our own analysis in
this chapter, as we move through his textual materials and the production
of his images shot by shot. In short, Christensen makes sure the audience

knows that it took three years to research and produce his visual thesis. As with the word and the face, this is stated abruptly for the benefit of context.

More title cards follow, filled with an authoritarian, first-person tenor. Lacking any established provenance for a voice-of-God tone that would only later become standard in the Griersonian documentary mode of the 1930s, Christensen takes it upon himself to invent this voice. The common suggestion that Luis Buñuel first generated this instrumentally impersonal tenor in *Land without Bread* (*Tierra Sin Pan*, 1933) is off by a full decade, ignoring the fact that silent films were anything but silent.[3] The director begins in this voice by establishing the witch as a chapter within a much longer constellation of practices, discourses, traditions, and institutions. This is empirically correct, as scholars from Gaston Maspero to Stuart Clark have emphasized in their own studies of the witch.[1] Among many others, Richard Kieckhefer has demonstrated how the long history of practical natural magic was enfolded into the specificity of European witchcraft in the late Middle Ages.[5] These findings have only taken root in the historical debates on witchcraft since the 1970s, which Christensen anticipates by some fifty years.

It is at this point in *Häxan* that Christensen gives us an image of the witch. It is a well-known woodcut that first appeared in Ulrich Molitor's *Von den Unholden oder Hexen* (1489), at the dawn of the witch hysteria in Europe, depicting two women feeding a boiling cauldron. Many of the stereotypical visual characteristics of the witch are not yet established: the age of the women is difficult to determine and they are far from the withered old crones we see later in Albrecht Dürer and Hans Baldung Grien.[6] Yet they are unmistakably witches. Their boiling brew evaporates into the air, appearing to cause a storm. Drawing on a trope that would instantly signify "the witch" from Shakespeare's *Macbeth* to the present, Christensen introduces the viewer to the subjects of his film via a classic example of the *maleficium* that people greatly feared from witches in the early modern period.

Christensen carefully limits what we can see of this image, narrowing the visible edges of the shot into a severe vertical line bisecting the screen. The shot is abrupt, barely onscreen for a few seconds before the intertitles return. Our focus is taken to the statement that primitive men "always" confront the inexplicable with tales of sorcery and evil spirits. This is obvious hyperbole, but not entirely out of step with the evolving scientific explanations of the time regarding the origins of human society. Echoing E. B. Tylor's argument that civilization always begins with the imaginative, superstitious responses of humans to a world they do not yet understand, Christensen then shifts to consider the power of belief.[7] *Häxan* at this stage appears to be aspiring to Max Müller's dream of presenting an objective, empirical "science of religion."[8]

Interestingly, the next image takes us to "imaginary creatures" thought to cause disease and pestilence in ancient Persia. A row of six human–animal hybrids confronts the viewer. Christensen immediately divulges his sources for this claim, citing Rawlinson[9] and Maspero[10] as authorities that trace the European belief in witches back to antiquity. Several shots of monstrous hybrid demons, drawn from Maspero, follow. Christensen is operating in a firmly rationalist mode here, linking these monsters to "naïve notions about the mystery of the universe" held by ancient people.

A re-creation of Egyptian astrological notions of the nature of the world immediately follows. This is the first explicit set to appear in *Häxan*, depicting (according to Maspero's information, the intertitle asserts) a world of high mountains, stars dangling from ropes, and a sky supported by strong pillars. A nameless assistant out of frame helpfully draws the viewer's attention to the important details.

As with any Universalist approach, Christensen traverses time quickly in the presentation of his thesis. No sooner have we glimpsed this scale model of the Egyptian cosmos than we are catapulted into the folklore of early modern Europe. Perhaps the singular feature of the witch craze in Europe is bluntly stated when Christensen informs us that the generalized evil spirits of ancient times are transformed into devils by the fourteenth century. Cutting from one to another, four iconic images of devils particular to the period flash across the screen, the film stock tinted an ominous, rusty red to heighten the effect.

These devils lived at the earth's core, Christensen tells us, with the earth believed to be a stationary sphere in space surrounded by layers of air and fire. Beyond the fire lay moving celestial bodies, ceaselessly rotating around the earth with the fixed stars far above and, "in the tenth crystal sphere," sits the Almighty and His angels, keeping the whole celestial system in mo-

"Christ in Limbo," from *Evangelicae historiae imagines*, after Bernardus Passerus (1593). Courtesy of British Museum, London.

tion. Intercut title cards offer explanation before Christensen helpfully reveals a working model of this cosmology, in this case drawn from Hartmann Schedel's *Liber Chronicarum*,[11] slowly pulling back the iris to reveal the medieval universe that he is has described. This moving representation of a terra-centered universe resembles the elaborate wonders found in Baroque *wunderkammer* meticulously assembled by the German elite at the time. It is an effective use of parallel editing to bring this lecture, delivered in text, to life in a visual manner.

While not explicitly designating the scene as such, Christensen is visually marking here what Frances Yates has called, after Cornelius Agrippa's handbook of the same name, "the occult philosophy" of the Renaissance.[12] The "rediscovery" of a large literature in Greek attributed to the name "Hermes Trimesgistus" by Marsilio Ficino (1433–1499), Giovanni Pico della Mirandola (1463–1494), Henry Cornelius Agrippa (1486–1535), and most powerfully of all, Giordano Bruno (1548–1600), galvanized a critique of the mainstream Church and represented a strong effort on the part of an elite group of scholars for an energized spirituality rooted in an Egyptian-derived wisdom handed down prior to that of the Old Testament. The core of Renaissance Hermeticism was a deep concern with astrology and the occult sciences, the secret essences of natural things, and the sympathetic magic that was made possible for those who mastered such essences and their relations to one another.[13] In short, the writings of Hermes Trimesgistus progressively provided the foundation for Ficino's relatively mild natural magic, Pico della Mirandola's Christian Cabalist, Agrippa's Christian magus, Tommaso Campanella's (1568–1639) utopian City of the Sun,[14] and eventually Bruno's full-blown Hermetic–Cabalist philosophy that sought through the power of astrology and magic to bypass the Church altogether, "operating" in such a way that the skilled magi could reach the Divine directly.

The fact that, starting with Ficino's Latin translations of the *Corpus Hermeticum*,[15] Renaissance Hermeticism, while not widespread or necessarily revolutionary, was based on a massive historical error in determining the provenance of the texts is relatively unimportant for Christensen's thesis here.[16] This does not blunt the historical accuracy of the connection Christensen is making between what was presumed to be "Egyptian" wisdom regarding the nature of the world and what he understood to be a religious tumult that arose out of this challenge in the decades preceding *his* witch. A full account of the rippling effects of the Hermetic–Cabalist tradition well exceeds Christensen's purpose here. It is clear, however, that the scrupulously mathematical astrology of Girolamo Cardano (1501–1576)[17] and the rigorously empirical studies of the natural world demanded by Bruno's attempts to operate as a magus paved the way for the science of Newton and Copernicus and for a new metaphysics to emerge that, over time, would come to be credited as the precursor to the modern scientific self.[18] The visual references to this tradition that flash by in *Häxan*'s first chapter argues in its own way for the central importance of this tradition well before scholars such as Yates and others[19] revolutionized our historical understanding of this tradition in the middle of the twentieth century. It is also relevant to Christensen's thesis that the violent refutation of the Renaissance magi was a crucial element in the battle with witches and Satan that demonologists and inquisitors took up in the sixteenth century. Jean Bodin's *De la démonomanie des sorciers* (1580) is an excellent example of the relation, as Bodin excoriates key Renaissance figures such as Mirandola and Agrippa as satanic precursors to the scourge of witchcraft that he felt was plaguing the faithful at the time. While often lost by modern scholars "in a hurry" to get to the details of witchcraft emblematic of demonological texts such as Bodin's, it is commonly the case that such anti-witch treatises begin with attacks on Renaissance magic and the Hermetic–Cabalist tradition that authorized it.[20] The stakes were quite high, as Giordano Bruno's execution demonstrated. In the sixteenth century, Hermetic magic and Cabalism became associated with the notion of "superstition" for Protestants and Catholics alike; for both reformers and counter-reformers by the close of the century, such superstition was a *crime*.[21] What appears elliptical in its presentation in *Häxan* is Christensen's reference to the twisted relationship between "Egyptian antiquity" and the witch trials in sixteenth-century Europe.

Christensen advances the analysis of visual culture in *Häxan* in the next scene, in an extended set piece devoted to the close examination of a miniature from the twelfth-century manuscript *Hortus deliciarum*. Tacking back and forth between the intertitle lecture and the careful consideration of details from the painting (again, with offscreen "pointers" to direct our gaze), the gory, elaborated reality of hell produced by the nun/artist Herrad of Landsberg jumps to life onscreen. It is clear that Christensen is overreaching by attributing a largely cohesive image of hell to a period when the nature of hell's location and "topography" was a subject of fierce theological debate. Although the early image corresponds to well-known later depictions of hell in literature (Dante) and art (Bosch) that filmgoers in 1922 could have reasonably been expected to know, Christensen's lecture strategically ignores debates and alternate conceptions of damnation that existed in the fifteenth and sixteenth centuries in Europe. For example, Sylvester Prierias's influential *De strigmagarum daemonique mirandis*[22] argued that fallen angels lived in the air, ungrounded and shapeless, manipulating the physicality of this air to act through witches and the wicked on earth.[23] While Christensen gives ample attention to the manipulation of air and environment later in the film (largely through the Wild Ride to the Sabbat), his authoritative approach is occasionally strained in close examinations such as the one he offers regarding the *Hortus deliciarum* miniature.

Despite the serious purpose Christensen asserts for *Häxan*, he appears to realize that spectacular moments of titillation are also necessary to carry through his visual thesis. This is evident in the following scene where a

Herrad of Landsberg, *Hortus deliciarum* miniature as it appears in *Häxan*, film still (Svensk Filmindustri, 1922).

working mechanical presentation of hell is revealed—the first real scene of movement in the film. The title cards suggest that Christensen "found" this visual machine dating to antiquity. It is unclear if the billowing smoke that at times completely obscures the actual mechanism onscreen (a rare technical misstep) is part of the workings of the apparatus or Christensen's own attempt to heighten the fiery terror of the scene. Either way, it works to amplify affect more than further analysis, leaving the title cards to offer a generic explanation that "during the Middle Ages, devils and hell were considered real and constantly feared."

After lingering on this spectacular hell device, *Häxan* returns to unfolding the necessary facts of witchcraft to the audience by moving to a shot of a woodcut depicting novice witches signing a pact with the Devil. Correctly following the procedural descriptions of famous witch-hunting texts such as Institoris and Sprenger's *Malleus Maleficarum*[24] and Nider's *Formicarius*,[25] Christensen emphasizes the essential agency of individuals in making a pact with Satan. The woodcut shows a novice witch being raised into the air by the Devil, foreshadowing the procedural reenactments that follow later in the film. Returning to his scholastic mode, Christensen notes that these images are drawn from Gustav Freytag's "A German Life in Pictures."[26] Two additional woodcuts from this source immediately follow; one depicting a witch milking an axe handle and the other showing a woman bewitching a man's shoe. While the ability to manipulate and pervert inanimate objects is empirically consistent with the historical record on witchcraft, Christensen offers no additional commentary or explanation of these bizarre images.

Christensen's visual narrative moves to the second essential element of medieval witchcraft: the Sabbat.[27] Molitor's *Female Witches Acting Together* woodcut (ca. 1493) is used to visually illustrate the gathering. Not lingering on the Sabbat itself, the film proceeds to witches acting in groups in order to cast spells on a cow, an entire village, and an unfortunate person who has fallen ill, drawn from Bourneville and Teinturier's *Le sabbat des sorciers.*[28]

Throughout this sequence of *maleficium*, Christensen emphasizes the arcane symbols used to ward off spells and the fact that the sick person is shown naked, a practice "that was habitual in the past." This vaguely salacious detail is emphasized further in the rapid succession of images finally revealing a group of nude women dancing around a demon at a Sabbat.

Though not explicitly stated as such in the title cards, a palpable sexual dimension has crept into Christensen's thesis. This element will be prominent throughout the film, depicted here in images of women "sneaking away" to attend Sabbats. Using woodcuts "passed on" to Christensen by "the French doctors Bourneville and Teinturier," the viewer is then taken through the typical rituals of a "secret satanic rite"—the abjuring of the Church by desecrating the cross, Satan renaming his initiates, and the horrific ceremonial banquet of the Sabbat. Christensen notes that the food for these banquets was often prepared using the corpses of executed prisoners (though in fact it was babies), continuing with an image found in Molitor, reproduced in *Le sabbat des sorcières.* Saving the best for last, witches (both male and female) are shown kissing Satan's anus as a sign of devotion. The image is a climax of sorts, and Christensen's textual accompaniment simply notes the activity without additional embellishment.

Molitor's *Female Witches Acting Together* as it appears in *Häxan*, film still (Svensk Filmindustri, 1922).

The Vicissitudes of Truth Telling in Early Cinema

Christensen's peculiar strategy in opening *Häxan* becomes apparent as the film progresses. Grounded in the realism of nascent nonfiction filmmaking, the director establishes his authority on the basis of citation. Inverting the typical structure of the monograph where the notes and sources would come last, *Häxan* visually grounds itself in citable evidence from the start. There are historical reasons for why this is done. Although hailed from nearly the moment of its invention as an instrument for "recording reality," the value of cinema as a vehicle for "telling the truth" about the world was increasingly regarded with suspicion by professional historians and social scientists in the early 1920s. Dominated by the "actuality" film, where the provenance of the images and the "historical tracks" of the observer were often obscured or even erased, purportedly nonfictional visual media were increasingly being judged inadequate to the tasks and protocols of the serious scholar at the time of *Häxan*'s release.[29] Michael Chanan has summarized this period in the history of nonfictional filmmaking as follows:

> When documentary was not yet documentary (but then fiction wasn't fiction yet either), when the medium was mute and each film ran only a minute or two, moving pictures hardly amounted to more than a miscellany of visual tidbits, which made no demands on literacy and thus spread easily and rapidly far and wide. The world on the screen exerted a magical attraction but remained anecdotal and predominantly iconic. In terms of public discourse, it was practically inarticulate, other than to reinforce already stereotypical images or create some new ones; in short, intensely fascinating but apparently ill-adapted to serving intelligent purposes.[30]

It is not as though scientists, journalists, and others devoted to making nature speak did not give filmmaking a try. In the waning years of the nineteenth century, anthropologists such as Alfred Cort Haddon, Walter Baldwin Spencer, and Frank Gillen were already using the new technology to fashion, with mixed success, proto-ethnographic films. Charles Urban founded the *Unseen World* series in 1903, merging the technologies of the microscope and the cinematograph to attempt to unlock the secrets of nature at its most minuscule level. Films such as *Attack on a China Mission Station* (1900), *Hunting Big Game in Africa* (1907), and *With Captain Scott, R.N., to the South Pole* (1912) sought to bring the immediacy of news headlines to life onscreen. State-sponsored war propaganda generated during the First World War, including *The Battle of the Somme* (1916) and *With Our Heroes at the Somme* (*Bei unseren Helden an der Somme*, 1917) mutated the desire to see far-off contemporary events through visual meaning-making machines that demanded not only attention but belief. The fact that these films nearly always made this demand by staging, as real, reenactments of purportedly real events only added to the early suspicion of cinema's ability to convey unvarnished, objective facts.[31] Even for films not surreptitiously staged, the reliance on actualities of iconic clichés, giving the viewer what they largely expected to see, proved to be a serious problem for those who wished to convey the complexity and depth of the world and of nature.[32]

The issue, widely discussed well before John Grierson's proclamation of the "documentary value" of Robert Flaherty's *Moana* in 1926, concerns the relation between a fragmentary visual artifact drawn "from life" and the truth value of any such fragments. Ultimately, this issue hinges on *mimesis*. What sorts of filmmaking practices can felicitously mimic life as such? Grierson's own elaboration of documentary recognizes this in asserting that the filmmaking form is the "creative treatment of actuality." Grierson was not the first to conceptualize the matter in this way, as Brian Winston shows that the Polish writer Boleslaw Matuszewski stated the issue in these same terms as early as 1898.[33] Crucially, mimesis was not only permissible for writers such as Matuszewski and early documentarians such as Edward Curtis; it was *indispensible* in the creation of valuable documentary works. Thus, a film such as Curtis's *In the Land of the War Canoes* (a.k.a. *In the Land of the Headhunters*, 1914) adhered to prevailing standards of expressing the real not despite its status as a reenactment but *because* of it. The truth of Kwakiutl (Kwagu'ł) life is evident through the *spirit* of Curtis's expert cinematic expression of what that life is, just as the reality of war was only truly evident to viewers through gaining a *sense* of the fighting as re-created in otherwise opposing accounts of the truth in the British and German Somme films.[34] Later accounts by film historians positing "fact" and "fiction" as oppositional binaries arising out of the earliest approaches to filmmaking were further exemplified by pitting the "realist" Lumière against the "fanciful" Méliès within a crypto-structuralist origin myth that falsely represented what "documentary" meant to pre-Griersonian filmmakers.[35] The "ahuman" witness of the camera is not enough, as this merely produces a blind sight that cannot, on its own, educate, enlighten, or even fully record the real in any ideal manner. This is not the first time that the gap between witnessing and the real has erupted in European history. As *Häxan* demonstrates, the question of evidence occupied inquisitors and theologians long before the invention of cinema.

Playing on the fact that, while the traces serving as evidence are quite different, a larger ontological issue binds them across the centuries, Christensen takes the unique tack of assuming the role of the *art historian* in this opening section of the film. This is a risky strategy, particularly given the static nature of the materials on display, but it does allow Christensen to shift the locus of the empirical to the materiality of images accepted as historical. Taking up this position in the opening chapter of *Häxan* also allows Christensen to have it both ways, in that he can simultaneously confront the viewer directly in the manner of an earlier cinema of attractions while also preparing viewers for the "diegetic absorption" that was coming to dominate the grammar of cinema in the 1920s.[36] Given the impossibility of filming witches several centuries "after the fact," *Häxan* creates present-day empirical images from artifacts of the time. Yet this analytic position does not guarantee that the images will be "brought to life" in any way. To the contrary, the vivisection of the historical image would tend to produce the same outcome that any vivisection would: death or deformity. Thus the risk, quite evident throughout the first chapter of the film, is that the presumed pastness of these images, their "deadness," will subvert the appearance of life that distinguishes cinema from other visual forms such as photography, painting, and printmaking. How well Christensen is able to elide this deadness is open to debate; undeniably, many viewers experience the opening minutes of the film as a plodding exhibition of "pictures of pictures." This reaction notwithstanding, the strategy of "reimaging" is methodological and intentional, an acknowledgment on Christensen's part that for a very long time "knowledge" in European terms consists first and foremost of "recitations of the known."[37]

While the opening chapter of *Häxan* may test the patience of the viewer, the logic of Christensen's visual strategy in this section becomes clearer as the film progresses. The director is laying a foundation for what comes next, though he is quite sensitive to the fact that a visual thesis demands a different relation to its sources. Thus, the parade of classic visual works in this opening section provides the ground not only for the arrangement of a thesis but also for the creation of *new* images, constituting its own evidence for what is at stake. Christensen accomplishes this by continually triangulating between paintings and woodcuts, photographs, and cinematic dramatization. This movement between formally distinct media at times more firmly aligns Christensen with those who affirm that "nonfiction" is a designation determined by techniques of *presentation* rather than simple *content*, including art historian Aby Warburg, filmmaker Chris Marker (particularly in reference to his famous 1962 "film of photographs," *La Jetée*), and the recent photography of Jeff Wall, Cindy Sherman, and Hiroshi Sugimoto, much more than with his own contemporaries in the cinema of the 1920s.[38] There are also echoes in *Häxan* of the creative displacements effected through Soviet montage and the use of fragments of found footage to assemble a singular work, with Esfir Shub's film *The Fall of the Romanov Dynasty* (1927) being the most obvious example.[39] *Häxan*, not having access to archival *footage* for obvious reasons, nevertheless re-presents the documents of the visual archive of the witch in a manner recalling the methods of Shub and other Soviet filmmakers such as Dziga Vertov. In formally similar films like Harun Farocki's *As You See* (*Wie man sieht*, 1986) and *Images of the World and the Inscription of War* (*Bilder der Welt und Inschrift des Krieges*, 1989), the "truth" gained by the reproduction of archival images is unlocked only through their *mobility* in the context of their new use.[40] As with Farocki, Christensen does not seek to embellish such visual artifacts in citing them, but rather *empties them out*, expressing through their preestablished frame a meaning that was hidden, resisted, or not even invented at the time of their origins. Understood in this way, the disconcerting effect of the opening chapter becomes more plausible, as *Häxan* disrupts what the audience can expect from the film. While the medium of expression is undoubtedly modern and allows for these uniquely moving images, the method Christensen deploys helps to cultivate a position that draws authority from an expertise based on the interweaving of the artistic and the scientific rather than an ideal "scientific self" premised on the polarization of the two.[41]

Visual Strategies: The Wild Ride

Two themes prefigured in this first chapter and foregrounded later in the film deserve treatment in terms of the visual strategies they employ: the Wild Ride and the hysteric. Our claims as to the methodological element of Christensen's image-making practices become clearer if we temporarily skip ahead to *Häxan*'s depiction of the violent moral disorder of the Wild Ride of the witches to their Sabbats. This scene appears in Chapter 4 of the film and is presented as a visual account of the old woman Maria's confession to the "crime" of witchcraft. We will fully analyze the density of this scene in the corresponding chapter of the book, but for now we will focus only on Christensen's complex use of works of art that originally appeared in fifteenth- and sixteenth-century texts by Hans Vintler and Johann Geiler[42] in the course of creating new cinematic images in *Häxan*.

Christensen's presentation of the Wild Ride is thrilling by any standard.[43] Making use of the special effects available to him at the time, the fury and terror of *Häxan*'s female wild riders stands out as one of many highlights of the film. By the early sixteenth century, the Wild Ride had become a standard element of both demonological and popular literary accounts of the activities of witches, folding older legends of wild hunters, the restless travels of the dead at night, and tales of the Furious Horde into the standardized script of the Ride. Particularly strong in what is today southern Germany and Switzerland, variations on the myth of the night people retained their durable immediacy deep into the twentieth century.[44] Charles Zika claims that in its various tellings the Furious Horde consisted of "cavalcades of demonic spirits and souls, especially of those who died before their time and enjoyed no peace—soldiers killed in battle, young children, victims of violent acts, and so on."[45] Folded into the exegesis of the ninth-century text *Canon Episcopi*, regarding the power of demonic illusion to deceive women into imagining that they could travel great distances at night, often in the company of the goddess Diana, the Wild Ride violently collapsed a multitude of characters and beliefs into a particular time and a singular image of the witch in sixteenth-century Europe. Christensen's own image of the Ride compels the same collapse, though one that assumes *fidelity to empirical evidence* in the time of the witch hunts. This is characteristic of *Häxan*'s cinematic *naturalism*.

There are many classic examples of images of the Furious Horde and the Wild Ride; two in particular stand out in relation to *Häxan*'s own visualization of the spectacular event. First is a clear correspondence between a woodcut from Hans Vintler's *Buch der Tugend* titled *Wild Riders on a Wolf, Goat, Boar, and Stool* (1486) and the special effect of Christensen's image of his witches flying through the air as part of Maria's confession in Chapter 4. This woodcut reflects its origins as a portrayal of Waldensian heresy (the subject of Vintler's text), depicting the riders, men, and vehicles as mostly animals.[46] While Christensen's image substitutes iconic objects such as brooms and cooking forks for beasts and reflects a discourse of the witch (found in Kramer in 1486) as being almost singularly female, it nevertheless takes direct inspiration from the classic woodcut in its perspective, its positioning of the riders in the frame, and the emphasis of the subjects that suppress depth of field against the void of an empty background. Vintler's woodcut, modified naturalistically to mirror the seemingly unnatural and impossible Wild Ride of the witch, *moves* in the film.

Christensen also modifies and brings to life characteristic representations of the Furious Horde, a supernatural band that was not originally associated with witchcraft at all. Again, this conjoining of witch image to demonological discourse reflects an empirically verifiable invention in the late medieval period and the Renaissance. In particular, Christensen's long shots of the witch's Sabbat, unfolding in the twisted chaos of the deep forest, recalls the woodcut *The Furious Horde* that appears in the 1516 version of Johann Geiler's *Die Emeis*. As with the echo of the Vintler woodcut in the Wild Ride, the perspective, framing, and composition of the image of the Sabbat in *Häxan* updates and transforms *The Furious Horde*, much as demonologists transformed the meaning of the Horde in the invention of the sixteenth-century witch. Again, Christensen is not only "inspired" by Geiler's image; he has in his creative activation of the image simultaneously produced an effect that corresponds to the empirical evidence of the witch's coming into being and exhibited what Charlie Kiel has termed "the oscillating value of the non-fictive."[47] Documentary elements can support, contradict, or even wholly become the narrative in early cinema; *Häxan* in this sense is consistent with other contemporary works in the oscillating value of its discrete artifacts.

Visually, *Häxan* offers innovation to the representation of demons that were commonly circulated in woodcuts, broadsheets, and paintings at the time. While the depiction of various lesser demons and fallen angels was quite common, they tended to be rendered as smaller versions of the horned Satan or as hybrid human–animal creatures with each "natural" species being traceable within the complete appearance of the demonic creature (such as in the Geiler woodcut just mentioned). *Häxan* does not simply *reproduce* these stereotypic images. Instead, Christensen at times broadens his regional frame of reference, drawing on works referring to witchcraft produced outside of German-speaking Europe such as Agostino Veneziano's painting *The Carcass* (ca. 1518–35) in relation to the Sabbat, or images that portray supernatural creatures that appear in negative sixteenth-century "guides" to pre-Christian Norse myth, particularly some of the woodcuts that accompany Olaus Magnus's *Historia de Gentibus Septentrionalbis* (in numerous printings from 1555), which appear to provide the inspiration for the "demonic children" Maria claims to have given birth to, revealed in her confession.

Agostino Veneziano, *The Carcass* (ca. 1520). Courtesy of Scottish National Gallery, Edinburgh.

Sabbat in *Häxan*, film still (Svensk Filmindustri, 1922).

Maria's confession in Chapter 4 of the film provides additional examples of the breadth of Christensen's visual assemblage of the witch and her activities. As with the discourse of the witch in the early modern period, figures from antiquity such as Saturn and Circe are also alluded to in the representation of the Sabbat in *Häxan*. In order to clarify our argument here, it is necessary to briefly analyze Christensen's composition of a series of brief shots in the Sabbat that refer to sixteenth-century representations of Circe and the link they made between the Roman goddess and witchcraft. In Maria's confession, Circe is indirectly named as "Satan's grandmother."[48]

Images associated with games of chance, gambling, tricks, slight of hand, and illusion were often part of Circe's repertoire. The logic here was that such games, seemingly minor performative elements of popular tricks and entertainment, were actually rooted in the same demonic power of illusion as more obvious forms of *maleficium*. Elements of Christensen's image here appear to be directly referring to a number of well-known visual representations of Circe in the sixteenth century, particularly a woodcut from the workshop of Michael Wolgemut and Wilhelm Pleydenwurff and tentatively attributed to Albrecht Dürer that appeared in the *Liber Chronicarum*, titled *Circe and Her Magical Arts Confronting Ulysses and His Transformed Companions* (1493). Although the literal confrontation depicted in this woodcut between Circe and her assistant on the shore and Ulysses and his companions on a boat is absent in *Häxan*, the flowing beauty of Circe herself is echoed in the film's image and the table cluttered with instruments of chance and magic directly corresponds to the association Christensen is intending to make here. Other surviving images from the time echo *Häxan*'s meaning here as well, albeit less directly. These would include the 1473 woodcut *Circe with Ulysses and His Men Transformed into Animals* from Giovanni Boccaccio's *Buch: Von den hochgeruemten frowen*, and the pen-and-ink drawing *The Children of Luna* from the *Housebook Master* or *Master of the Amsterdam Cabinet* (1480). The hybrid animal–human forms of the demons dancing around the "grandmother," the woman's surprisingly young and beautiful appearance, the wind whipping around her, her elevated position in frame as if she is floating in the air (she is actually positioned on a ledge, but this is very difficult to discern until the "grandmother" is shown in medium shot, entering a door), and the array of objects and instruments she wields all

Giovanni Benedetto Castiglione, *Circe Changing Ulysses's Men into Beasts* (c.1650). Courtesy of British Museum, London.

Close-up of Circe's instruments in *Häxan*, film still (Svensk Filmindustri, 1922).

point to the refiguring of the mythological figure of Circe as a powerful witch in the service of Satan.[49]

Visual Strategies: Hysteria

A key component of *Häxan*'s thesis is that the power of the witch is reanimated in modern times through the signature of hysteria—something foreshadowed in the first section of the film, and reinforced in Christensen's complex strategy of tacking between painting, photography, and moving cinematic image in *Häxan*. The "historical framing" in the *longue durée* of his thesis in the first chapter is carried throughout the entire film. There are a number of scenes in *Häxan* that activate unconscious associations in the viewer between melancholia, witchcraft, and possession. For example, while none of these paintings is explicitly displayed in *Häxan*, Christensen appears to have taken direct inspiration for a number of his shots from Lucas Cranach the Elder's famous *Melencholia* series of paintings. Produced between 1528 and 1533, these four paintings that depict the supernatural environment haunting a female melancholic bear many similarities with elements that Christensen brings to life in *Häxan*, including the Wild Ride, terrifyingly unnatural children, and a general sense of sexual and societal disorder swirling around a placid, passive female protagonist.[50] It makes sense that Christensen would evoke Cranach as the paintings reflect an empirical strain of the discourse of the witch that highlighted the susceptibility of the melancholic to the Devil's illusions and hence to witchcraft and especially possession. In the *Melencholia* series Cranach composes the face of his female subject as a *mask*, the swirl of activity around her signifying what lay behind her placid, deceptively beautiful façade.

Interestingly, there are several points in *Häxan* where Christensen self-consciously composes similar faces, simultaneously concealing and revealing the turmoil that lay behind them. In particular, later in the film we find Brother John's troubled reverie in the face of his repressed, possessing desire for the Young Maiden and the mask/face of the unnamed hysteric that is the subject of most of the film's concluding chapter. In both cases, Christensen draws a link between these carefully framed faces and *possession*, a mobile element moving between the pact of the witch and the obsessed state of the hysteric. Reversing Aby Warburg's assertion that donning a mask constitutes an active attempt "to wrest something magical from nature through the transformation of the person," Christensen's figures invert this polarity by appearing to be worn by the mask.[51] Thus, the re-membered face of Cranach's melancholic in these shots works as a relay between Christensen's moving images of the witch/hysteric and unseen, but obviously present, iconic images of Charcot's hysterics. This is entirely consistent with Charcot's belief that artistic works of demonic possession and melancholia were reliable evidence of hidden and misdiagnosed mental disease. As Avital Ronell has put it, "The scientific imperative, the demand in the nineteenth century for an epistemological reliable inquiry in the nature of things, derives part of its strength from the powerful competition represented by fascination for the freak and the occult, which is always on the way to technology."[52]

By formally constructing "the witch" through a cinematic iteration of *metoposcopic* naturalism, Christensen could not agree more. Although left unsaid in the opening chapter, the imperative Ronell cites is progressively etched on the face of the images the director produces, be they explicitly "photographed" icons or evoked as echoes and memories. Using a strategy similar to that famously deployed by his closest filmmaker contemporary, Carl Theodor Dreyer, Christensen will build from the elements of this opening chapter to a complex, expressive interplay of *face* and *tableau* in order to bring the witch to life in *Häxan*.

What Is *This Thing?*

> An image is strong not because it is brutal or fantastic—but because the association of ideas is distant and right.
>
> Pierre Reverdy, "L'image" (1918)

The first chapter of *Häxan* draws to an abrupt close, its tone descending from the overwhelming affective force of images of explicitly sexual acts with Satan. Christensen actively avoids taxing the audience with any further explanation or lecture. We find images of witches flying (this time "returning home" after a "merry dance") as a final set of title cards blandly state that

Lucas Cranach the Elder, *Die Melancholie* (*Melancholy*) (1528). Courtesy of Scottish National Gallery, Edinburgh.

Brother John in *Häxan*, film still (Svensk Filmindustri, 1922).

images such as the ones the audience has just seen "are often found on famous witch Sabbath pictures from the Middle Ages and the Renaissance."
Three more dense images then flash in secession (it is unclear if they show
Sabbats, hell, or some combination of these on earth) and then a final title card
appears and is held several beats longer than those immediately preceding it,
having the effect of a door held for a moment before slamming close this
chapter of the film.

Häxan gets off to an undeniably peculiar start. In our view this is due to
the formal, methodological ambition of the work, particularly in regard to
the conscious triangulation of ontologically distinct image-objects arising
out of paintings/woodcuts, photographs, and cinema. Christensen is trying to make the power of the witch real in a way that seems impossible
through a film. Invading the domains of the human sciences, particularly
those of the art historian and the ethnographer, Christensen will not remain
content to faithfully reproduce traces of the past, devoting the remainder

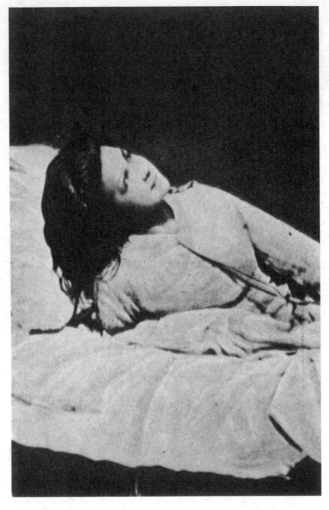

Augustine, "Attitudes passionelles: The Call," from Désiré-Magloire Bourneville, Paul Régnard, Jean-Martin Charcot, and Édouard Delessert, *Iconographie photographique de la Salpêtrière: Service de M. Charcot*, 3 vols. (Paris: Progrès Médical, 1877–80).

of *Häxan* to willing a new life into texts and images. The director's "atlas of images at work" strategy is strikingly reminiscent of the methodological innovations of Aby Warburg, particularly in relation to Warburg's unfinished *Mnemosyne* project.[53] It is worth quoting Philippe Alain-Michaud's summary of Warburg's scheme at length:

Hysteric hears the call in *Häxan*, film still (Svensk Filmindustri, 1922).

In *Mnemosyne*, photographic reproduction is not merely illustrative but a general plastic medium to which all figures are reduced before being arranged in the space of a panel. In this way, the viewer participates in two successive transformations of the original material: different types of objects (paintings, reliefs, drawings, architecture, living beings) are unified through photography before being arranged on the panel stretched with black cloth. The panel is in turn rephotographed in order to create a unique image, which will be inserted into a series intended to take the form of a book. The atlas, then, does not limit itself to describing the migrations of images through the history of representation: it reproduces them. In this sense, it is based on a cinematic mode of thought, one that, by using figures, aims at not articulating meaning but at producing effects.[54]

Heightened by the effect in cinema that *everything in frame appears to be alive*, the strategy will prove to rupture the very perceptions of "deadness" or "pastness" that allows the modern viewer to evade the power of the witch that Christensen will forcefully assert is still with us. The time of the witch, in all its multiplicity and exigency, will be brought out of the past and into the present by appearing to register the form of life itself on film.[55] Thus, the alienating distance of both the objects presented in *Häxan*'s first chapter and the characters they refer to is necessary to begin with, as the task of the film now becomes the closing of this distance between the two-dimensional surfaces of photographs and celluloid and the three-dimensional sense of lived experience. Similar to Robert Wiene's *The Cabinet of Dr. Caligari* (1920) in this respect, *Häxan* shows an affinity with the Cubist art contemporary to its release in the tension it strategically heightens by ignoring or contravening the perceptive "rules" of formally distinct image artifacts.[56] In later decades, artists and filmmakers such as Gerhard Richter (*Atlas*, 2006) and Jean-Luc Godard (*Histoire[s] du cinema*, 1988–98) have taken up Warburg/Christensen's methodological logic in their own attempts to link the dimensions of the image with life. Within the arc of this movement in Christensen's film, the objective knowledge of witchcraft is opened to the perception of otherness in the witch, the demonologist, the hysteric, and ultimately the scientist by way of a visible unity of the senses unique to the director's method.

"The ethnographic surrealist," wrote James Clifford, "unlike either the typical art critic or anthropologist of the [1920s], delights in cultural impurities and disturbing syncretisms."[57] We are not claiming that *Häxan* is ethnographic in its formal approach, yet Clifford's description does echo the links we are drawing here between radical approaches to the image in art and subversive methods deployed in documenting the real that were roughly contemporary to the film.[58] The transgressive approach to the archive, to classification, and to expression that the film exhibits also is akin to methods deployed in the journal *Documents* (1929–30) nearly a decade later. Edited by Georges Bataille, *Documents* willfully transgressed institutional genres through its "subversive, nearly anarchic documentary attitude," an attitude that Christensen plainly shared.[59] What distinguished *Documents* from Warburg's *Mnemosyne* and *Häxan* is that the former seizes clichéd objects and then systematically empties them out in the course of its own expressions. Bataille and his contributors sought to defamiliarize the clichés, disturbing the placidly deceptive surface of the mundane in their fragmentary, juxtaposing methods of critique and presentation. In contrast, Warburg and Christensen begin by collecting mythological, figurative givens seemingly quite distant from the "really" real. Starting at radically different places, the outcomes of these projects converge on the same nodal point— unsettling distances between myth and the everyday that in turn produce expressive works that are themselves quite unsettling. It is obvious in light of this shared methodological aspiration why the surrealists would take in-spiration from *Häxan*, brazenly (and unfairly) advocating Christensen over Dreyer as the Scandinavian filmmaker of note in the 1920s.[60]

David Bordwell groups *Häxan*, along with Carl Theodor Dreyer's *Leaves from Satan's Book* (*Blade af Satans Bog*, 1921), Maurice Tourneur's *Woman* (1918), and Fritz Lang's *Destiny* (*Der müde Tod*, 1921) within a tradition of "episode films" in the classical period of silent cinema.[61] This is consistent with our argument regarding Christensen's film, as all of these cinematic works weave together episodic fragments in order to draw parallels and cor-respondences across situations and characters. More explicitly than the others, however, *Häxan* also deploys the techniques associated with War-burg's *Mnemosyne* and Bataille's *Documents* for purposes of affectively em-phasizing the dark, chaotic forces that lurk under the smooth surface of the everyday. The parallels Christensen draws are therefore not simply between characters or situations but across domains of sense that cut across time. Thus, the episodic structure of *Häxan* not only allows characters seemingly out of a dead past to live again, it also draws the phenomenology of the hysteric and the work's own contemporary time to the surface. Shadowed by the specter of an everyday fractured by mechanized global war, *Häxan* in turn brings its witches, inquisitors, and hysterics alive in the haunted now of the film's reception.[62]

In short, *Häxan* is *promiscuous*. It is neither wholly artistic nor scien-tific. It aspires to seize a quality Ulrich Baer granted only to photography when he wrote, "Films fail to fascinate in the same way as photographs do, because they invite the viewer to speculate on the future—even when irresistibly tempted to do so—only on the level of plot or formal arrange-ment. Photographs compel the imagination because they remain radically open-ended."[63]

Häxan calls Baer's assertion into question. The opening chapter does not offer a speculation as to the future. It disorients the viewer, leaving her with the insistent, fundamental question, "What *is* this thing?" It compresses times past and future into a sequence of clichéd images that traverses the steep slope between past and future in the form of an event. This is not a plot. Rather, it is a strategy to "compel" the viewer, although we would not limit this compulsion to the imagination alone. In other words, the inability to automatically categorize *Häxan* emerges out of a formal strategy rooted in an *epistemic virtue*. In science, such virtues demand that the subject know the world and not necessarily the self; *Häxan*'s demand is greater in its own way as it demands *both*.[64] Thus, while Christensen never backs away from his claim that *Häxan* offers a truthful examination of the witch that can stand up to the test, he also deploys strategies of evidence making that would have been familiar to the subjects of his film. As Joseph Leo Koerner puts it: "In the later Middle Ages, in practices ranging from persecuting witchcraft to meditating on Christ, techniques were developed to draw distinctions among visual phenomena, differentiating, say, physical objects from fantasies, dreams, and diabolical or artful deceptions. Some of the best testimonies of this sorting operation come from artists. This is not surprising given that image-makers specialized in manipulating one thing (their materials) in order that a viewer should see something else."[65]

While Christensen's materials might have been radically different than those of an artist in the late Middle Ages, his aim to manipulate these materials in order to make something invisible visible is consistent with his aims. This description, of course, could also be applied to experimental scientific techniques without much alteration to the stated aims of tests taken under the signature of such disciplines.

For Christensen, objective knowledge itself has been possessed by the uncanny, rendering "imagination" or "reason" alone inadequate to bringing the witch to life, to forcing her to speak to what is already known in her pathological language of diabolic proofs. The witch must be experienced in her own milieu, a satanic biome that we will presently argue is one that Christensen represents as her state in nature. As it moves from the first chapter to the second, *Häxan* constitutes an extension from the techniques and virtues of *Mnemosyne* to those of the *nature film*. In other words, the first chapter of *Häxan* is the presentation of a series of clichés—visual clichés and stereotypes of the witch, fragments which were most likely already familiar to the viewer. This is hardly a waste of time, however, as these clichés (what Deleuze terms *figurative givens*) will not only provide the empirical evidence for Christensen's thesis but will also provide media from which the director will conjure the power of the witch. It is important to note that Deleuze discusses figurative givens in reference to painting, not cinema; thus, the concept would not seem to readily apply here.[66] Yet we suggest that Christensen is attempting to do something quite paradoxical, which is to release the movement of the painting and the woodcut through the cinematic image. Indeed, as we move through the film, we cumulatively gain the *sense* that *Häxan* is a *living tableau*. This is by no means an accident. The film excels in providing the ground for this sense, possessing the spectator through the immediacy regardless of whether the viewer logically knows that the represented event is already in the distance. This quality sets *Häxan* apart.[67]

Evidence, Second Movement: Tableaux and Faces

> What was normal [by the seventeenth century], at all levels from the patrician to the plebeian, was the marriage of word and image. That "mutuality" may be tacit: thus heroic painting in the grand manner does not generally carry with it, in so many words, a commentary, keying in the mythological figures; yet artists habitually gave their paintings titles, mottoes, tags and quotations, and their works abound in literary allusions. But very commonly the interleaving of the verbal and visual is quite explicit.
>
> —ROY PORTER, "Seeing the Past" (1988)

> The originary world is therefore both radical beginning and absolute end; and finally it links the one to the other, it puts the one into the other, according to a law which is that of the steepest slope. It is thus a world of a very special kind of violence (in certain respects, it is the radical evil); but it has the merit of causing an originary image of time to rise, with the beginning, the end, and the slope, all the cruelty of Chronos. This is naturalism.
>
> —GILLES DELEUZE, *Cinema 1* (1986)

An Unnatural Business

What better place to begin pictorially than in the underground lair of an old witch named Karna? A hoary, wrinkled woman moves busily in her dark underground room, the scene cluttered with a variety of objects that are difficult to recognize and yet generate an ominous, dreadful sense that something malevolent is going on here. The old woman tends to a pot over the hearth in the middle of the large, dank room; not precisely the cauldron central to the witch stereotype, but certainly close enough. An accomplice, a somewhat younger version of the sorceress, enters the room with a large bundle of straw, roughly tossing the bundle to the side. The bundle falls heavily and the old shrew quickly reveals why; hidden within the shock of straw is a corpse (or part of a corpse), exposed by the old woman parting the straw and drawing a lifeless hand from within the bundle. The meaning of her previous dialogue ("Tonight the stars shine favorably over the gallows hill") is answered by this gesture.

Complaining about the quality of the item her coconspirator has procured ("Ugh! What a stench!"), the old woman nevertheless proceeds to examine the hand carefully, suddenly snapping one of the fingers off the decaying hand. Noting that the finger of the thief may be "too dried out" to lend any power to her brew, Karna nevertheless ties a string to it and lowers it into a large cask. Her partner does not respond and blithely stirs the small cauldron boiling on the hearth. In a series of flowing, intercut close and medium shots, Christensen reveals the terrible ingredients of the cauldron—ingredients that (save for a large, writhing snake and a still-alive toad feebly attempting an escape) are unidentifiable in their strangeness. This mildly creepy reveal visualizes a stock cliché of the witch stereotype regarding the ingredients she uses: of nature and yet revoltingly unnatural and unwholesome at the same time. The casual suggestion of cannibalism also references the popular understanding of the witch, something we discuss in relation to the witch's Sabbat later in the book.

Someone approaches the entrance to the old woman's home. She is hidden and nearly frantic to enter before she is seen. A customer. Karna shows the plain, middle-aged woman in. Warily surveying the scene, the customer gets directly to business. She is in need of a love potion to be used to entice "a pious man of the church." Karna, being a savvy entrepreneur, has a range of choices to offer her customer. In succession, she offers a potion of "cat feces and dove hearts boiled in the moonlight" or a stronger brew rendered from "a young and playful male sparrow." As Karna speaks, the customer visualizes the potential outcomes of each potion, Christensen dissolving back and forth from our perspective eavesdropping on the transaction to scenes of the customer administering the potions and their results, first one of ardor, then one of frenzied sexual arousal.

Although played for amusement, this sequence serves the purpose of presenting several additional elements of the witch that were not precisely germane to the stern lecture in the opening chapter. First, the conflation and merging of myriad forms of popular magic with the *person* of the witch is plainly shown here. Classic witch-hunting manuals such as the *Malleus* sought to associate common magical practices of protection and healing with the witch; this association had to be strongly asserted by elite writers of the time as the connection was not obvious and was never fully absorbed into popular discourses of what constituted a witch.[1] Second, the stereotype of a debased and corrupted priest is forcefully introduced at this juncture. The object of the customer's affections, surrounded by wealth and comfort, is a fat, uncouth, and (if one is to "read" the face) vaguely stupid friar who appears to be completely subjugated by his desires. In each short fantasy aside, the friar is shown eating lavishly prepared food in the manner of "an animal," dismissive of the woman serving him prior to administering of the potion— he is then instantaneously and completely overtaken by his desire once the witch's concoction is ingested. The question that forms in the mind of the viewer is not about the effectiveness of the potion, but why the customer would desire such a slovenly, corrupt man in the first place.[2]

Christensen leaves this particular question hanging, although his insultingly satirical portrayal of this gluttonous, lustful priest has an antecedent in the strategically vulgar aspects of Protestant discourse against the Catholic Church in the sixteenth century. No less than Luther himself was known to slander, scandalize, and offend his adversaries using explicitly vulgar language against them. Taking cues, artists of the period extended the instrumentalization of slander through the production of proto-pornographic images of bishops, priests, and the pope engaged in myriad obscene acts. It was not uncommon for the pope to be depicted as the Antichrist in such works. Ignoring Thomas Aquinas's assertion that scandal, either "active" or "passive," is always a sin, Protestant propagandists sought to offend, using explicit rudeness as a weapon in the battles of the Reformation and Counter-Reformation. For them, satirical, injurious images were regarded as expressions of empirical realities; this was not in spite of their obscenity, but rather *because of it*.[3]

As empirically grounded as *Häxan's* depiction of this persuasive discursive form is, Christensen's fidelity to historical facts did not endear his film to censors. The difficulty Christensen had in getting *Häxan* released in Germany serves as a good example of the issues raised in nearly every country outside

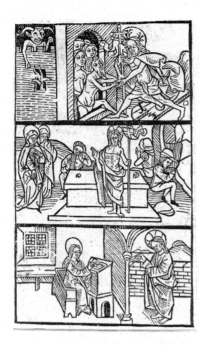

Childhood of Christ; the Passion; three
cuttings from *Itinerarium beatae Mariae
virginis*, printed by Johann Reger, Ulm.
Courtesy of British Museum, London.

of Scandinavia where distributors attempted to show the film. Although
several private screenings prior to its public première generated praise for
Häxan in the German press, the film's release was denied when first submit-
ted to German censors in February 1924. Only after substantial cuts was the
film publicly shown in Berlin in June of the same year. The controversy over
Häxan lingered despite Christensen substantially truncating the work. In
January 1925, the Prussian Ministry of the Interior formally petitioned
the censors to rescind their approval, citing scenes such as the one described
above as expressly intending to offend religious feeling, to threaten public
order, and to "brutalize."[4] Although the petition was rejected, such efforts
severely restricted showings of *Häxan*, laying the groundwork for its unfortunate status as an "unseen classic." Interestingly, it was quite often scenes
such as those depicting the emotional states and desires of the clergy that
aroused as much ire as scenes showing nudity, demonic violence, sex, and
cannibalism.

   Christensen ends this lusty introductory scene with Karna, not yet taking her customer's money, suggesting that the best remedy of all may be her

Devils Watch while a Jesuit Sodomizes a Young Woman, from *Historische Print en Dicht-Tafereelen, van Jan Baptist Girard, en Juffrou Maria Catharina Cadiere* (1735). Courtesy of Bayerische Staatsbibliothek, Munich.

salve. This ointment (which she dramatically displays by carving out a portion with a blade) is so powerful that the "pious monk might directly come to [her] chamber." She notes that they would fly together at night in amorous bliss. Again, Christensen introduces another core element of the witch stereotype: the salve. More than any other substances, magical salves and

Desirous priest in *Häxan*, film still (Svensk Filmindustri, 1922).

ointments were considered particularly powerful and troubling by demonologists, and their use was a sure sign that an individual was engaged in
witchcraft.[5] *Häxan* stops before showing whether Karna's customer takes her
up on the offer.

Häxan's thesis is now unfolding. Without warning, the film cuts from the
scenes with Karna to a moody, melodramatic scene with a young man hurrying another corpse through the streets of the village, finally arriving at a
door where he is ushered in to the home with his load. Joined by a second
young man, the two of them laboriously heave the bundled body down into
the basement, at first unnoticed, observed by a woman awakened by the commotion. Through a series of cuts between the men and the spying woman,
the female corpse is revealed. They are troubled and nervous. "Listen
brother, shouldn't we pray . . . ?" asks one as the other stands dramatically
poised with a large knife, ready to plunge it into the cadaver. Pray they do,
beseeching the Holy Mother for forgiveness for cutting open a body in order to learn its secrets. The spying woman, overcome with curiosity, walks

in on the macabre scene. Realizing what is about to take place, she runs screaming into the streets, shrieking of the desecration brought about by "two witches."

This scene is quite dramatic, but also puzzling in its placement in this section of the film. Certainly the evolving practice of anatomical dissection and the development of pathological anatomy as a central element of medicine drew objections and accusations of blasphemy and desecration (although in an historically uneven way); what is strange is that this was not a strong element of the witch stereotype in the time period to which *Häxan* ostensibly restricts itself. Dissection and dismemberment of bodies did exist as a religious practice, but tended to be restricted to the securing of holy relics from the bodies of recent dead regarded to be "saintly" or in public displays for the purpose of understanding anatomy and general curiosity.[6] One must wait nearly two centuries before the trope of the "body snatcher" in the name of anatomical science really comes into being. While inquisitors and demonologists were certainly concerned with abuses visited on corpses, the evidence to which Christensen cleaves in the inquisitorial manuals drafted during the early modern period simply does not support the idea that anatomists or physicians were mistaken for witches. It is clear that Christensen wants practices associated with developing and testing medical knowledge to enter his visual narrative from the earliest moments, even at the risk of contradicting the historical record.

Christensen's inclusion of this event, directly following the scene with the lay practitioner of natural magic, is peculiar, as the film clearly depicts Karna as a "witch" while implying that the young men are simply "mistaken" for witches. We can only speculate as to the reasons for this empirical lapse, but the short scene does allow for a strong visual correspondence between misunderstandings of illness and scientific medical practice and the reasons for witchcraft accusations. *Häxan*, for all its cinematic license and unspoken complexity in relation to the power of the witch, never explicitly moves away from its self-positioning as a *scientific* investigation. As such, the idea that we are waiting to discover the "real" reasons for witches is never far from its agenda, whatever else *Häxan* actually communicates to its audience. To undercut the idea that Karna is a witch, however, would have severely weakened her scenes, and Christensen's tone here is one of unambiguously positioning her as not only a misguided healer but also someone who explicitly believes she *is* a witch and acts more or less in the conspiratorial manner that is described in the *Malleus* and elsewhere. Yet also aspiring to have the audience feel the power of the witch, Christensen must constantly pull back from the dramatic outcomes of what *Häxan* depicts. Thus, pulled between these tensions, we see a dramatic slippage in the scene of the two amateur anatomists. Unable to fully negotiate the multiple demands Christensen himself makes of the film, we are presented with a dramatically useful error in the presentation of the evidence for his thesis.

Christensen notes at this point (via title cards) that it was common for everyday people at the time to see witchcraft as the source for a wide variety of misfortunes; in order to illustrate this statement, he moves to a scene of a conflict on the street at night between an old woman sleeping on some steps and a passer-by. The man angrily rousts the woman, accusing her of "bewitch[ing] the legs of honest people." Her aggressive reply is to bewitch the man's jaw, forcing his "filthy mouth [to] remain open for eternity." The stricken man, a look of terror on his face, collapses on the very steps from which he has rousted the old woman. This seemingly throwaway scene is interesting in that Christensen's depiction does not appear to illustrate the paranoid delusions of witch-crazed villagers; rather, it is a scene that unambiguously shows the bewitching of a man. While the audience can speculate as to potential somatic or psychosomatic causes for the man's sudden affliction, *Häxan* unequivocally plays the scene as a demonstration of the power of the witch. Unlike the previous scene of misunderstanding arising from the activities of the amateur anatomists, there is no visible cause for the action here except the angry spell of the old woman.

The function of this short scene is subtle. As the film unfolds, it is clear that Christensen is aware that the witch comes into being in overlapping vectors between learned discourse and popular belief. This requires some ground for agreement between the two domains; one clear aspect of this overlap is the widely held set of beliefs regarding the "nature" or "essence" of women. Echoing Christina Larner's assertion that witch trials were gender-related but not by definition gender-specific, the rhythmic alternation in this chapter of *Häxan* between women and men acting according to their supposed natures provides some insight into not only why the accusations against women came to dominate the witch trials but also how it was possible, under some conditions, for the accusation to be leveled at men as well.[7]

The gender that Christensen invokes here, although perhaps somewhat overdetermined in its reliance on functionalist binaries and contrarieties, is not far from the social–functionalist explanations offered by anthropologists such as E. E. Evans-Pritchard, and more recently by historians such as Stuart Clark who have been inspired by anthropological attempts to apprehend the figure of the witch.[8] Returning to Larner's insight, it important to note that *Häxan* does not represent the vicious pursuit of witches as a straightforward instance of misogyny on the part of the Church or sixteenth-century civil authorities. Rather, it was the case that women often came under suspicion of being witches because they were understood to be particularly susceptible to lust, avarice, and jealousy by their very nature. Women, acting in accordance with this presumed nature (as the mean old woman briefly demonstrates), were therefore "naturally" susceptible to witchcraft. Men could also be witches, but generally only in instances where they were clearly acting against norms. The young proto-anatomists, therefore, have given in to a curiosity that was by the standards of the time morbid and unnatural. They, too, are taken to be witches, but only because what they are doing is *not* expected of them, a sure sign of unnatural forces at work. By contrast, women could signal the mobilization of these same unnatural forces simply by enacting elements of the nature that they were always already presumed to possess.[9] This difference will be important as *Häxan* progresses through its thesis.

The uneasy structural ambiguities carry over into the following scene. "So it happens with witchcraft as with the Devil; people's belief in him was so strong that he became real." This intertitle follows the line Christensen established earlier regarding the error and false consciousness of those who believed in witches and generates a nagging, almost unconscious, reluctance about Satan being made real. This phrasing is consistent with the positions taken by Clark, Roper, and a host of other recent historians that one cannot approach witchcraft or possession from a vantage point in the present without granting some legitimate status to the ways in which the Devil and witches were not only asserted to be real but were experienced as such.[10]

Reassuring Visions

Häxan's visual expression of this point constitutes one of the best-known scenes in the film. A priest, revealed later to be Father Henrik, engaged in intense prayer, is suddenly confronted by the Devil himself. Seeming to emerge directly out of the large Bible the priest is reading (in fact, popping up from behind the stand supporting the book), Satan is monstrously intimidating, leering at the terrified priest who backs away in horror. Played by Christensen himself, Satan taunts the priest and his colleague, who has rushed over in aid. Spreading his terrible claws over the pages of scripture, Satan dominates in even the holiest places (a church) and through things (sacred text), a fact that Christensen renders powerfully in this sequence.

This scene is campy by today's standards, in part due to the lasciviousness of Christensen's Satan and the hysterically overwrought reactions of the

Satan appears in *Häxan*, film still (Svensk Filmindustri, 1922).

Satan appears in *Häxan*, film still (Svensk Filmindustri, 1922).

harassed priest. Its power to shock, however, remains intact. Satan erupting forth for the first time during an act of prayer, in a church, and confronting a pious believer and instrument of God, visually conveys the terrifying and reassuring sense of power and threat Satan possessed. Although often understood as the point in which *Häxan* begins to slide into the territory of farcical reenactment, this short scene is among the most empirically consistent sequences in the entire film, particularly in relation to an understanding of the power of the witch and the Devil and the sense of these beings that existed in relation to life at the time.

Prior to Satan's dramatic entrance, we see the religious trappings, but there is no evidence of God's acknowledgment or answer to the friar's prayer. Yet, despite the obvious shock of the event, Father Henrik's confrontation with Satan at the very moment he beseeches God is shown to consolidate, rather than dissipate, his pious resolve. This resolve in *Häxan* reverberates outward toward the implied (visual) term in this powerful image; Satan is indeed right in front of Father Henrik but the witch is not far behind. Encircled by these diabolic figures, the friar can ironically perceive the truth of the words he had just been carefully reading. On its own, the Bible is unable to convey information or simply *communicate* in a reliable, testable manner.[11] Coming face to face with Satan, at the very moment of the Word's perception in the mind of Father Henrik, the required supplementary proof is given via the concrete, threatening body of the evil one himself. Body and Word conjoined in Christensen's cinematic image; the ritual the scene began with can now *speak*.

The sense that Satan could be everywhere, positioned just out of sight, pulling the strings of his demonic human puppets is not limited to the early modern period. Nor is *Häxan* the only film where the notion appears. The figure of Satan as the power behind calamitous events in human history is also used in Dreyer's *Leaves from Satan's Book*. Like *Häxan*, *Leaves* is an episode film, but in Dreyer's slightly earlier work, Satan's *malefic* presence not only traverses far spaces but also crisscrosses time itself in what Bordwell has termed "a density of parallelisms."[12] Composed of four sections showing the crucifixion of Jesus, the Spanish Inquisition, the French Revolution, and the then-current civil war in Finland, *Leaves* portrays Satan as being potentially behind all calamity, driven to subvert humankind as punishment for his rebellion against God. In Dreyer's version, the Devil is awesomely powerful but also somewhat pitiable in that he can do nothing else but disrupt and destroy as an enforced condition of God's punishment. This rendering of Satan, at times strangely sympathetic, relies explicitly on discourses of theodicy and God's ultimate permission for Satan's deeds, debates reaching back well into the fifteenth and sixteenth centuries. Importantly, Dreyer figures Satan as a problem of the present in the last episode of the film, with the Bolsheviks violently engaged in revolutionary struggle being the manifestation of a transcendental demonic power. There is a great deal of debate as to Dreyer's political leanings in associating Satan with contemporary communists. The disagreement about whether Dreyer was a "conservative" or engaged in explicitly ideological filmmaking with *Leaves* is important on its own terms, but is not relevant here.[13] It is, however, important to note that Satan was a very powerful figural vehicle in Scandinavian cinema at this time and that Christensen's own rendering of the Devil as potentially "being everywhere" would have certainly reverberated with contemporary debates over politics and evil, and with other films such as *Leaves* that take up a similar expressive strategy. As in the time of Thomas Aquinas, this question of theodicy ultimately cannot help but come around to God's seeming absence from the world; in the wake of the devastation of the First World War, this question for many had never been more pressing.

Several scenes building on the displayed power to deceive and bedevil human beings follow the shock of the Devil's first appearance in *Häxan*. In sequence, Satan's ubiquity and lustfulness is shown over three scenes: (1) terrifying a woman lying in bed at night (it is unclear whether this is a nightmare); (2) enticing a nude female somnambulist out of her home into the forest, where she eventually kneels before a demon who embraces her; (3) appearing at the window of a woman sleeping with her husband and violently "encouraging" her to come with him—she does not immediately succumb, but is shown in several close-up shots grimacing and licking her lips, then in medium shot arching her body in erotic pleasure, and finally returns the Devil's embrace in her bed. While the charged eroticism of this sequence is played for shock and titillation to some degree, it remains consistent with the strategy of giving a powerful sense of the descriptions of Satan that exist in the demonological literature to which *Häxan* refers, particularly regarding the materiality of the Devil and sexual encounters with him. As Walter Stephens points out, proving sexual relations with the Devil was an essential task in many witch trials and served as crucial empirical evidence for Satan's existence.[14]

It is notable that in this series the targets of poisonous attention are young, beautiful women. While they are described as the "Devil's companions," these scenes are ambiguous as to the will and agency of the young women who become entangled with Satan's erotic power. Moving to the next sequence, *Häxan*'s narrative shifts focus to a character that actively seeks the companionship and assistance of the Devil—the viewer is brought back to the figure of an old woman.

Venusberg

Häxan now returns to the basement lair where this chapter of the film began. It is apparent that the set is the same one inhabited by Karna earlier, but this time the viewer is introduced to Apelone. Christensen's first-person form of address in the intertitles is jarring as he directly queries Apelone by asking if it is "from the eternal fright of the pyre that you get drunk every night, you poor old woman of the Middle Ages." As the audience "hears" this condescending question directed at the old woman, she is shown shuffling and stumbling around the dimly lit basement, although interestingly she is not shown to be actually drinking. Christensen is forcing the issue somewhat, as there is an obvious distance between what is "said" and what is seen. Sarcastically driving a further wedge, doubt as to the director's felicity arises at various stages in *Häxan*, with this short sequence being a prime example. While it is unclear if Christensen intended to make a self-consciously critical statement, this quasi-humanist magnanimity reveals what Catherine Russell has termed "condescension toward the Other."[15] Her critical analysis of Buñuel's *Land without Bread* is also appropriate here: "Surrealist ethnography might therefore be a means of denoting the strategic roles of ambivalence, cruelty and empathy in refiguring the ethnographic relationship in postcolonial culture. Buñuel evokes the dangers of the photographic image and its implicit historical structure, marking the deep divide between those 'out there' in the real, and those who watch 'in here,' in the auditorium."[16]

Christensen is hardly a surrealist *ethnographer*, but his dreamlike *historiography* of witches is presented with the same double signature, simultaneously relying on and then disavowing the authority of science and its humanist social iterations. The unsteady trace of this signature become all too apparent in harsh moments such as the director's address to Apelone in *Häxan*, revealing Christensen's documentary to exist as an often cruel secondary revision of what is, or was, already in the world. The correspondence between art and science driving this revision is left unmarked, but, echoing Walter Benjamin, it is quite clear that Christensen's art sets out to "conquer meaning."[17] This move, made manifest at the expense of a defenseless figure from the past, scarcely distinguishes *Häxan* from the science it overtly aspires to or its contemporary approach to the unfortunate hysterics that become the focus of the final two chapters of the film. In this sense, *Häxan* is rigorously consistent throughout.

Apelone falls into a stupor in the corner; the Devil appears. In the course of rousting Apelone, Christensen's Satan performs some of the most outlandish and obscene gestures of the film. In particular, his frenzied thrusting

Satan "churns his butter" in *Häxan*, film still (Svensk Filmindustri, 1922).

of what appears to be a butter churn positioned suggestively between his legs unmistakably is meant to intimate masturbation. Christensen's wild, onanistic gestures may draw shocked laughs today, but it is an effective, purposeful performance, producing a disturbed affect that ratifies the blasphemous, obscene experience of being confronted by Satan himself. The shouting, tongue-wagging, powerfully stroking Devil is truly disturbing and grotesquely attractive in the way that only a night vision can be.

Apelone, like the audience, appears terrified when she comes to realize what she is seeing. She is nevertheless compelled to follow where Satan commands her to go. Satan suddenly flies up in the air and out through a high window; now fully aware, the old woman hurries over to the window and is hurled into the air herself. The film cuts to a shot of a castle, noting to the audience that this is Apelone's "dream castle" (with the potential double meaning of the word "dream" left open to interpretation) and the place where the Devil will fulfill her wishes.

Häxan then cuts to a close-up of an unconscious Apelone. Interestingly, although she is shown flying through the air immediately prior, the ride itself and her mode of conveyance (typically represented as a broomstick or chair in woodcuts at the time) are not shown. A full visualization of the infamous Wild Ride will have to wait until later in the film. We see gold coins pouring down on Apelone's head, awakening her. The shot widens to show the floor of the well-appointed room where she is now covered in these coins. Stunned, she excitedly gropes the coins, unaware that Satan is watching her. She dumps a pile of coins on the table located in the foreground of the room when suddenly the coins begin to fly up in the air, disappearing. Apelone is alarmed and feebly attempts to corral the vanishing coins. She fails miserably and pleadingly looks up into the ether where the coins have vanished. *Häxan* then moves to the door of the room, now open. The coins remaining on the floor in front of the door fly away as well. A close-up of Apelone's frantic, greedy face betrays her confused terror to the viewer. The

Apelone's disappearing coins in *Häxan*, film still (Svensk Filmindustri, 1922).

coins completely fly away from Apelone as she chases them, stumbling clumsily into the next room.

The effect of the disappearing coins, generated by running the original shot backward, is nearly as old as cinema itself. The Lumière brothers' short actuality, *Demolition of a Wall* (*Démolition d'un mur*, 1896) is believed to the first film to deploy this technique by simply running the film backward through the projector, showing first a wall being torn down and then the smashed wall miraculously reconstituting itself. If the conceptualization of time as an "arrow" dominated thinking in the late nineteenth century, then the revolutionary potential of this simple technological reversal is obvious. These unnatural reversals were in centuries past attributed to the Devil's deceitful manipulation of natural laws or the senses, so it is no stretch to suggest that cinema's early association with magic is a logical one, an association linked to the special effect in early horror such as J. Searle Dawley's version of *Frankenstein* (1910). Mary Ann Doane attributes this correspondence to the "semiosis of cinema's own technological condition," whereby such conditions are transformed into "legible signs."[18] By attempting to bring the witch to life, such signs serve as both subject and subtext. The Vitagraph/Edison film *The Artist's Dilemma* (1901) is more explicitly a precursor to *Häxan* in this respect. The short film begins with an artist carefully painting a model in what appears to be a Victorian drawing room. As described by Doane, a "clown/demon" emerges from the clock and proceeds to "unpaint" the original picture and with rough brushstrokes substitutes his own photo-likeness version of the model, which he then proceeds to bring to life and help down off the canvas. Although used in a much more sophisticated way in *Häxan*, the reverse-motion effect in *The Artist's Dilemma* bluntly demonstrates the technique's general purpose in both films. As Doane writes about the earlier short, "The parallel between the realistic portrait and the film image—both inhabit a frame and emerge out of blackness—demonstrates that [*The Artist's Dilemma*] seeks to reinscribe the uncanny likeness of the cinematic image as magic, and magic as the underside of science."[19]

Over two decades later, *Häxan* would deploy the same special effect in its scene of Apalone's visit to Venusberg, despite the technique having gone out of favor with filmmakers as a gimmick of earlier cinema. In this case, the technique's connection to what was even in 1922 a somewhat anachronistic cinematic past grounds its use within the subject of the film, allowing it to work precisely *because* of its association with the past. The alterity of the antiquated method here edges the radical power of the witch a little closer to the viewer.

In the next scene, a sumptuous feast is laid out for Apelone. Having forgotten her lost fortune in gold as quickly as it appeared, Apelone greedily moves over to the table to eat. Before she can even begin, a small demon claws his way through a nearby door, tearing through it sharply. Thinking better of the situation, Apelone backs off and escapes the room through another exit, entering a dark room dominated by a large wall painted with Satan's face. The eyes fix their gaze on the old woman, glowing, as the door (positioned as his nose and mouth) opens to reveal a group of witches dancing wildly in a circular fashion, darting in and out of sight through the orifice. These glimpses of the unfolding Sabbat are intercut with close-ups of Apelone's beseeching face, tears streaming down her cheeks. Apelone charges the door, but it slams shut, Satan's glowing eyes ludicrously crossing as his gaze continues to hold the old woman. On the side of the room a beautiful maiden beckons for Apelone to join her, a slightly older woman observing from immediately behind. We see Apelone's relieved smile, but before she can move, the scene abruptly changes; the old woman starts awake, back in her dark basement, the moon shining through the open window. We now see a man, slumped in a chair, holding a trumpet. He moves toward the gauzy light of what appears to be dawn filtering through his window and blows the trumpet. Other trumpeters answer the call in the twilight. Apelone, in profile, stares out the window toward the sound of the echoing horns.

Through Apelone's night visit to Brocken, Christensen has broadly introduced two more elements of the witch stereotype: the Wild Ride and the legend of "Venusberg" as a gathering place for the Sabbat. These elements return throughout the film. As the Wild Ride itself is not shown in any great detail in Christensen's depiction of Apelone, our discussion here will follow *Häxan*'s rhythm and will come to the Ride later. We can at this point, however, say more about the settings of Brocken and Venusberg.

Apelone's travel to "Brocken" is either a concession or an error on Christensen's part. While the place may have been recognizable to viewers in 1922 through the famous scene of the Sabbat at Brocken in Goethe's *Faust*, the seventeenth-century writings of Johannes Prätorius, and Christopher

Marlowe's attack against conjurors (and Jews) in his circa 1593 play *Doctor Faustus*,[20] the setting would have been completely unknown to the people depicted in *Häxan*. The best-known meeting place for the Sabbat was the Heuberg ("Hay Mountain") in southwestern Germany. Sometimes also called "Venusberg," this remote site was suffused with myth well before the emergence of the witch in the late Middle Ages. Believed to be the peak where the goddess Venus convened her clandestine court, the Heuberg was known far beyond its local region, as evidenced by Nider's mention of the place where witches assembled at the Council of Basel in 1435 and the fact that the site is directly named in trial transcripts from the 1520s. Even in the case of the rare male accused, Chonrad Stoeckhlin, well-known today through Wolfgang Behringer's careful study of the case, the Heuberg was named and played a prominent role in the mobilization of the witch stereotype.[21] Of course, the fact that Institoris gained, by serving as an inquisitor, the "ethnological" experience with witches that provided the basis for his writing of the *Malleus Maleficarum* in this region is also quite significant, given the inspiration Christensen took from the book.[22]

Dreamtime

Apelone's night flight in *Häxan* serves to place the audience in the milieu of "dreamtime" of the sixteenth century as described so clearly by Behringer.[23] The presentation of rigorous empirical details pertaining to this dreamtime will come later in the film. Christensen is well aware that the time of the witch craze will appear naïve, strange, and distant to a viewer in 1922; his task at this stage of the film is to decisively close this distance. The task is trickier than it seems, particularly given Christensen's own indecision. On the one hand, setting the mood of the film through the context of Apelone's apparent hallucination serves the need of affectively positioning the audience for what follows; this is hardly revolutionary. Yet the steep slope of Christensen's naturalism in *Häxan* edges into view. Recalling Deleuze's description of naturalism in cinema, at this point in the film we do not yet have a proper sense of either the witch or the demonic source of her power. What we do know, or more precisely what we can *sense*, is the originary world from

which these figures come. Apelone making a quick trip to the fantastic Heuberg or Satan lasciviously mocking the Church through the startling deni-
gration of its servants are essential elements of what *Häxan* aspires to, but it
is only a beginning. Now building on top of the repertoire of images shown
in the first chapter, Christensen is visually rendering oral and written sources
as elements of the image. This is parallel, but is not identical to, what the
original artists were doing in creating the woodcuts and paintings seen earlier. Christensen is now working *cinematically*, seeking to create affective
conditions that differ from those of painting or drawing. Again, this difference, this naturalism that is now properly cinematic, relates to *time*. Thus,
we cannot *yet* see the witch (Apelone hardly appears to qualify as one), but
without Christensen's efforts to affectively shift the sensory world of the
viewer as he does in this chapter she would effectively remain invisible to
us. This was not the situation for Dürer, Baldung, or Cranach. Artists at
the dawn of the Reformation sought to represent the void as a figure;[24] Christensen, veering away somewhat from the obvious Protestant influence on his
art, seeks to coax the figure of the witch out of this profound void.

In a film as indebted to painting as *Häxan*, the impulse to ascribe to it a
label such as "Romantic" or "Expressionist" is great. Such a move is not
entirely without merit. The Romantic oscillation between the macabre and
the lyrical appears to be one obvious correspondence. The attention paid
to the extremes of mundane life and the strategies by which Expressionist
painting sought to externalize states of mind do at times appears to be another. Read in this way, *Häxan* can be understood as being very similar in
its modeling to the other great masterpiece of early horror cinema released
in 1922: F. W. Murnau's *Nosferatu* (*Nosferatu, eine Symphonie des Grauens*).
Like Christensen, Murnau took direct inspiration from the visual art of centuries past, composing his scenes in a manner that reflected the influences
of a diverse set of painters, including Arnold Böcklin, Giorgio De Chirico,
and Caspar David Friedrich.[25] While some film scholars have challenged the
usefulness of the label "German Expressionist cinema," it is nevertheless undeniable that Murnau and other German directors such as Wiene drew
heavily from the visual strategies and creative energies of works associated
with these movements in painting.[26] Considering the close ties that existed
between Svensk Filmindustri and the German studios—not to mention

Christensen's own strong connection to this film industry—the idea that *Häxan* can be placed among films such as *Nosferatu, The Cabinet of Dr. Caligari*, and *Faust* (F. W. Murnau, 1926) is plausible.

While we acknowledge the reasonableness of this grouping, our own analysis demonstrates that one should be wary of *Häxan* fitting neatly into given categories. *Häxan* does not play well with others. Rather than defending or refuting questions of categorization, our own approach has been to emphasize how *Häxan* corresponds with a variety of traditions without seeking to assimilate the film fully within one over another. It is true that we claim that *Häxan* exists as a naturalist film, but this claim is intended to mark a relation rather than a rule.

Faces, Tableaux

In the context of our refusal to wholly associate *Häxan* with any single designation, our concern here shifts from whether *Häxan* is indeed an Expressionist film to how *Häxan* expressively operates in relation to other films contemporary to it. Specifically, we return to questions of tableau and face, as it is along these two poles that one can discern similarities with some (Dreyer) and differences with others (Murnau). Bordwell has identified these two elements as essential to understanding Dreyer's early films, particularly *Mikaël* (1924), and while the interplay of these two elements produces a much darker outcome in *Häxan*, they are nevertheless similar.[27] In particular, the stillness and fixity of tableau-like shot composition that is evident in the works of both filmmakers and distinguishes them from nearly all their contemporaries. Implying a closed system in such shots, the affect is often one of a suffocating organization. Dreyer takes this principle to new heights in films such as *Master of the House* (*Du skal ære din hustru*, 1925) and particularly *The Passion of Joan of Arc* (*La passion de Jeanne d'Arc*, 1928). Christensen himself had already deployed an evolving version of this logic in his previous films *The Mysterious X* (*Det hemmelighedsfulde X*, 1914) and especially in *Blind Justice* (*Hævnens nat*, 1916). *Häxan*, too, puts this principle to work, albeit in correspondence with very specific countershots of great mobility and freedom. Crucially, Christensen ruptures the tableau element of *Häxan* in order to visually express the lively, mobile power of Satan; specific ex-

amples include the bewitched priest chasing his servant, Satan's initial erup-
tion before the praying friar (both in Chapter 2), and of course the extended
scenes of the Sabbat (Chapter 4) and the possession of the nuns in the convent (Chapter 6). Thus, only the upsurge of Satan's power can break the im-
mobility of the tableau, which we find well into the sequences regarding
possession and hysteria. This visual strategy bears a precise relation to how
demonologists conceptualized the workings of the Devil's power in practical terms.

The face in *Häxan* also disturbs the tableau element of the shot. As with
the tableau, the film resembles but is not identical to Expressionist art or
cinema in this specific regard. Deleuze summarized Expressionism as the
play of light and darkness, with the mixture of the two producing an effect
that suggests either "fall[ing] into the black hole or ascend[ing] towards the
light."[28] In Deleuze's analysis, the face concentrates this series, elevating
what may be symbolically rendered as "light" or "dark" to a power or a quality.[29] In Dreyer, the viewer finds that the face allows for a perspective that,
in its suppression of depth of field and backgrounds generally, makes this
affective power *mobile* along the lines of Deleuze's meaning: *mobile* as *spiritual* in its effect. Although perhaps not as finely developed as in Dreyer's later
films, this formal characteristic aligns *Häxan* with Dreyer's work in the 1920s
through to *Vampyr* (1932) and serves to distinguish Christensen's use of
tableaux from that of Murnau.[30] In Murnau the tableau frees the viewer for
introspection regarding nature in a kind of emotional, spiritual release. The
close-up is almost never deployed in many of Murnau's German films, as it
would structurally disrupt the affect he was seeking in works such as *Nosferatu*; compare this with the disruptive pathos the close-ups generate in the
German director's *The Last Laugh* (*Der letzte Mann*, 1924). In *Häxan* the tableau grounds the uncontrollable forces at work on the faces of those confronted by the power of the witch, constituting an intensive rather than
introspective power in the shot.[31]

The functional interplay between tableaux and faces in the course of
grounding *Häxan* in a naturalist cinema of the demonic is only faintly reg-
istered at this specific point in the film. At the conclusion of this chapter,
the audience is comforted by the thought that this is *all* a delusion, not yet
realizing that they have not simply witnessed Apelone's "hallucination," but
are in fact themselves being drawn into someone else's dream. Living among

the restless dead and in the specter of an increasingly active evil, the dream-beings for Apelone would not have presented themselves as the harmless images of an overactive subconscious; there is no question of what is "real" in her dreamtime. Setting us up in this way, it is increasingly clear in *Häxan* that the dreamtime of the witch is not as harmless or as distant to the modern viewer as we would like to believe. We are not yet fully held by the witch at this point in *Häxan*, but she is starting to move closer.

Häxan presents itself in the formal procedure of a progressive unfolding of the material world through the style of a lecture. This form would have been quite familiar to those members of an educated, literate public in 1922 interested in the apprehension of the world through humankind's chief instrument: *science*. A hierarchy emerges, and thus art and religion are subjected to the scrutiny of scientific proof. Drawing force from the near-messianic belief in the perfectibility of man, *Häxan*'s opening chapter invites viewers into a narrative of the witch and an exploration of the wonders and "errors" of the past. But Christensen is not simply addressing an assembly of experts; rather, he is trying to draw a spectating public into the zone of the witch. Then, as now, a filmed lecture would not generally qualify as a satisfying film-going experience for anyone but the most dogmatic viewer. Continuing in this way may not even qualify *Häxan* as a work of cinema.

Häxan thus abandons the rather overbearing didacticism of the opening by moving directly into the "underground home of a sorceress in the year of the Lord 1488." Now the static images presented earlier come to life on the screen. The full force of this reanimation only becomes apparent as the film moves forward—*Häxan* gives an affective form to the otherwise abstract, myth-like notion that witches were widely believed to be real and powerful in the early modern period.

Christensen is composing images of a figure that is already present. The witch of the opening chapter was a perceived thing held at a distance, a set of ready-made circuits of recognition and association that nearly any viewer would instantly recognize. From here on out, *Häxan* progressively moves away from the clichéd figure, bringing us dangerously close to the *real* power of the witch in the process. Our use of the hazardous term "real" is meant quite precisely, as *Häxan* proves to be a film based on a magnified form of *realist* cinema. More specifically, *Häxan* is rooted in a *naturalist* impulse. Through an assemblage of fragments from this basis in its formation of a cumulative image of the witch, *Häxan* will, in Deleuze's words, make apparent "the invisible lines which divide up the real, which dislocate modes of behavior and objects, are supercharged, filled out and extended."[32] Christensen not only marks out an image of the present in *Häxan* but also conjures the aura of a seemingly timeless origin myth, collapsing the distance between them in the violent multiplicity of *Häxan*'s surfaces and figures. Christensen's witch is not only here *now*, it has always been here: a figure of nature. Demonologists of the fifteenth and sixteenth centuries explicitly associated their work with the "advancement" of natural knowledge.[33] In a deliberately perverse style, Christensen appears to agree with them. It is a testament to Christensen's skill that the viewer is seized by the witch despite the nagging suspicion that one should "know better." Like everyone else, the viewer is gradually ensnared.

Terms like "capture" or "seizure" denote physicality, a manual activity that lay at the heart of Christensen's method. This seems paradoxical considering the virtual nature of images, but it is important to remember where *Häxan* begins—with woodcuts, drawings, and paintings that originate from an act of touch. While it is questionable if the cinematic image can ever achieve the tactility of the painting or woodcut, Christensen aspires to surpass the commonsense division between the tactile and the optical in order to generate for cinema viewers what Deleuze in a different context called a "haptic vision."[34] This corresponds precisely to the tactic in the opening chapter of the film, to present figurative, clichéd givens as they establish the ground by which Christensen can transform these figures from virtual givens to haptic modifications through the remainder of the film. The eye and the hand work together in *Häxan* to mold the image. This is possibly due to Christensen's ability, working expertly with cinematographer Johan Ankerstjerne and set designer Richard Louw, to correspond the movement of cinema in line with what are more commonly painterly images. In this sense, *Häxan* bears a close relation to the silent films of Carl Theodor Dreyer in that the "plane-ness" of the image, the negation or perversion of depth of field (particularly through the close-ups of faces), and the occasional use of eccentric, disorienting continuity editing (eye-line mismatches, violation of the 180-degree rule, etc.) produces a molded, affective, tactile quality that compels the viewer to grope for the image.[35] In the case of *Häxan*, the "clay" from which Christensen will mold these haptic images—the tactile substances by which one can touch (or be touched by) what is happening onscreen—has already been literally shown to the audience and will continue on occasion to recur throughout the remainder of the film. *Häxan* seizes (the audience) and is seized (by the witch), establishing a formal cinematic strategy that parallels the very problems of seeing and touching *virtual beings* (such as devils) that transformed what constituted evidence for the presences of the witch.

In Chapter 1, *Häxan* began with image fragments, narratively held together through an almost belligerent narrative logic. In Chapter 2, Christensen empties out and "paints over" these figurative givens in his own assemblage of the witch. In moving to the live action of "the underground lair of a sorceress," the film is now utilizing the repertoire of oral tradition: "old witches tales." Now firmly in the creative mode of composing images on top of given visual surfaces, *Häxan* will move progressively from *tales* to *theology*, and finally, *diagnosis*. It is a rigorously logical structure that works simultaneously to throw the fissures of the real into relief while also boldly expressing the tangible singularity of the power of the witch.

The Viral Character of the Witch

In witchcraft, words wage war. Anyone talking about it is a belligerent.
—JEANNE FAVRET-SAADA, *Deadly Words* (1980)

In the third chapter of *Häxan*, Christensen returns to his didactic mode of presentation. Through Franz Heinemann's *Rites and Rights in the German Past*,[1] he presents to the viewer a common investigative technique deployed by inquisitors and witch hunters: trial by water. Heinemann's image is intercut with another image, a detail of a bound, naked woman undergoing a similar trial, which Christensen tells us is drawn from Eduard Fuchs's *Illustrated Social History from the Middle Ages the Present*.[2] The title card explains the scene: "If she floats, she will be pulled up and burned. If she sinks, the judges thank God for her innocence."[3]

The scene is not a "trial" but rather an experiment. It needs to be clear, however, that the aim is not to identify "the seed" or origin of a particular evil, only evidence of its presence. As we will show in this chapter, such trials operate though a form of *non-knowledge* that comes from the mastery of *nonsense*. In lieu of direct, unmotivated proof, it was common for inquisitors to undertake such experiments in order to obtain evidence. These procedures, when understood as adjacent to witnessing, display the growing concern felt by authorities at the time regarding their ability to *prove* what they *knew* in advance to be true. Experiments such as "trial by water" demonstrate not an indifference to the truth and a retreat into superstition, but rather a deep (if not misguided) appreciation of cause-and-effect relations, as well as the forces at work in the natural world. Christensen presents this experimental practice as barbaric, but he also allows its logic to remain clear. After all, only a witch could manipulate the properties of a natural substance like water in order to avoid drowning in this experiment. These practices concerned *cases* suggestive of and empirically linked to *general laws*. The case said something about the world—and once a case was established, it would spread like a contagion.

Making a Case

Despite the powerful illustrative quality of the case study, there is an obvious aporia opened when the individual provides "evidence" of something different from what is understood as a collective phenomenon. In practice, the individual is useful insofar as she represents an ideal type.[4] The particular necessity of the individual to perform this role is already problematic, as psychoanalysts and anthropologists can certainly attest. In nonfiction cinema, the distance between the individual case and the ideal type is nearly impossible to close, clearly reflected in the famously controversial proto-ethnographic film also released in 1922, Robert Flaherty's *Nanook of the North*. Aspiring to document the life of the Inuit through the reenacted case study of Nanook (real name: Allakariallak) and his "family" (also actors, of sorts), Flaherty's film expresses a *certain* truth regarding the total social environment. There is great irony in the fact that Flaherty can do this only by undercutting his claim to objectivity, as the film affectively brings case study examples alive as *subjects*. At the level of a cinematic artifact, this effect does not by definition suggest failure, as the force of Nanook's life not only provides empirical evidence as to his mode of living but also allows for a reflection on "nature," "humanness," and "modernity" rooted in the haptic qualities of Flaherty's images.[5]

Although more fragmented as a work, the middle chapters of *Häxan* move in a direction similar to Flaherty's film through the presentation of a single reenacted witch trial. No specific, historically documented trial is mentioned, suggesting that *Häxan* will present for the viewer a "typical" procedure. It is a case study populated by ideal types drawn from *similar* facts and events.

Häxan also enters the case study through a scene of illness and diagnosis. In light of the film's strategy to present the audience with a typical event, such an entry remains true to its purpose, as concrete forms of witchcraft were found in the real misfortunes that people faced, particularly maladies with poorly understood causes and cures. In such times, the *maleficium* of the witch was palpable and its physicality highlighted throughout *Häxan*. Although many other forms of *maleficium* are cited across primary historical sources (destructive weather magic, assault of farm animals, etc.), by far the most profoundly *embodied* (and cinematic) type would be sickness and unexplained death.

Hiding in Plain Sight

The scene of diagnosis in *Häxan* is beautifully composed, a tableau of figures expertly arranged. A man (Jesper the Printer) lies in his sickbed, motionless, as a male healer (Peter Vitta) and various female members of the household attend to him. Christensen's framing moves between medium shots and close-ups of the attendants, not revealing who they are, but establishing that the man's wife and infant are among those in attendance. Their faces are tense and apprehensive. Scenes of the activity in the bedroom are interspersed with scenes of activity in an adjoining kitchen. Older female servants bustle and silently carry out their tasks. In the next room, a tiny old woman slips into the kitchen unnoticed by the servants, hurrying across the frame. "The power of lead will soon reveal it," Anna, wife of the Printer, is reassured. The male healer moves to a small cauldron at the foot of the bed and then carries out the simple rite, which ends with the reading of melted lead suddenly solidified in cold water. The power of Saturn is invoked in the ritual. The inscrutable physical result of the lead experiment is revealed: the cause of the malady is "atrocious witchcraft."

The direct reference to Saturn in the ritual is significant to the arc of Christensen's cinematic strategy, as it provides the basis for a chain of possible associations. Although we will offer a fuller explanation of the ancient Roman god's link to concepts of witchcraft later in the book, it is important to note that Christensen is using a well-established trope when identifying Peter Vitter as one of "Saturn's children" in this scene. Georg Pencz's woodcut *Saturn and His Children* (1531) captures the character of the belief that Saturn serves as patron to social outliers, including the poor, elderly, and disabled, as well as criminals, Jews, cannibals, magicians, and witches.[6] Throughout the century, Saturn's mythological violence was increasingly associated with the demonological violence of Satan. Witches were understood to be clients of the Devil's patronage much as Saturn's children were bound to the ancient god and to one another.

In short, what results is a conjoined figure of Saturn/Satan.[7] This conflation serves as additional evidence of natural magic, illustrating how long-established techniques of practical magic were subsumed into the figure of the witch. Yet, while this is hinted at in the scene of Jesper the Printer's diagnosis, it is important to remember that Peter Vitter is *not* suspected of being a witch in *Häxan*. It was certainly not unheard of for lay healers and diviners to find themselves among the accused, but this is not Vitter's fate.[8] In fact, once he has completed his diagnosis, he disappears from the film; interestingly (or perhaps tragically), so does the sick man, Jesper the Printer.

This scene in *Häxan* is significant for its depiction of the steady medicalization of witchcraft, as well as for offering us clues to the director's own position in his cinematic thesis. Specifically, while it is difficult to determine Christensen's intent, the scene draws a clear line between the empirical instruments and techniques of the healer and the persistent notion that Satan himself was, in fact, *the* principal authority of the natural world. Henri Boguet offers one of the clearest examples of this belief in his *Discours des sorciers* (1610), arguing to the point of dogmatism that Satan was not only fully subject to the laws of nature but was also the foremost master of the knowledge of natural properties and the techniques of their instrumentalization.[9] If Christensen's only intent was to provide yet another visual demonstration of people's naïveté and misguidedness during the early modern period, he did not have to go to the lengths of this otherwise minor scene. Michel de Certeau observed that *every* exercise of trained judgment is authorized through the dark, ratifying force of theology.[10] The association between Peter Vitter and Saturn gives the veiled sense that the healer's powers

Georg Pencz, *Saturn and His Children* (1531). Courtesy of British Museum, London.

may originate, either knowingly or unknowingly, from dark forces that rat-ify *all* forms of natural expertise. As *Häxan* moves from theology to diag-nostics, it is difficult to avoid the thought that the witch's power may be, at its core, *identical* to that of the inquisitors, exorcists, and physicians who seek to "remedy" her acts.

His task accomplished, Peter Vitter packs up and readies to depart through the kitchen. The Young Maiden,[11] the sister of the sick man's wife, Anna, is foregrounded, shown slowly turning the lump of lead over in her hands, ner-vously fondling the oblique object. The Young Maiden then rushes to Peter Vitter demanding to know the identity of the sorceress. He answers crypti-cally, "You might see that witch, before you wish to . . ." The Young Maiden repeats these last four words, wandering through the kitchen in a daze. The scene returns to the old woman who entered the kitchen earlier.

Peter Vitter is not simply playing a cruel game with the Young Maiden. He has, in effect, done everything possible and expected. As a man who has a certain "touch" when it comes to divining the causes of supernatural mis-fortune, he has identified the mechanism of the Printer's malady. This is not, however, the same thing as having knowledge of the specific source. Confirming the suspicion of witchcraft is, if anything, a form of *non-knowledge* derived from a mastery of *nonsense*; it opens a gap in knowing (specifically, *who*). It would be imprudent and dangerous to go further. It is not up to Peter Vitter to name the witch; this burden is on the Young Maiden and the rest of Jesper's family. Still, it begs the question, why would Peter Vitter stop short of giving a name?

In fact, Peter Vitter *has* given a name; he has announced that a witch is at work here, and this naming is essential to the arc of counteracting the *maleficium* that has been visited on the household. The act of naming is dan-gerous enough for Peter Vitter, as he has not only laid hands on Jesper but has also come in contact with the (un)natural force that has stricken him. The healer knows that the source of evil is close—hence his cryptic message to the Young Maiden as he departs. He is, we assume, divulging everything he knows in order to turn the family in the right direction, simulta-neously assigning their roles within the progressive structural logic of witchcraft. Who is close enough to Jesper the Printer to undertake this evil spell? Who would have cause to do so? These are questions that *the family* must answer. Peter Vitter has given them more than enough information for them to carry out their tasks; it is now up to them to move to the next step.[12]

Peter Vitter has departed. The Young Maiden stares wide-eyed, tears streaking her face. Although it would seem that we are not at a particularly riveting point of the narrative, the jolt of the image effectively mirrors the Young Maiden's state of mind. This is also the first of many shots in *Häxan* where Christensen employs an extreme close-up of a face. The Young Maiden instinctively utters a "sacramental" phrase and identifies the old woman: Maria the Weaver. It is clear that it is not going to take long for the Young Maiden to put into further motion what Peter Vitter began.

Maria begs for a meal and the Young Maiden reluctantly agrees. The disdain for the uncouth old woman is evident on her face. She serves Maria, betraying a suspicious look as the woman ravenously eats. Christensen returns to tight facial close-ups to convey Maria's animalistic mannerisms and the Maiden's horror. The images move from close-ups of Maria's food smeared face, to a medium shot of the Young Maiden disgustedly turning away, again to Maria, and finally to the Maiden turning to face Maria, her face contorted with revulsion. The Young Maiden runs into the adjoining bedroom where her sick husband rests, slamming the door shut behind her.

The dynamic, tensile relation between tableau and face has reached its height, and continues from this point forward in *Häxan*. As we discussed earlier, this relation between "tableaux" and "faces" is established from the beginning of the film. As he moves into his case studies, where clichés and particularities collide, the dynamic qualities of Christensen's approach begin to overwrite the figural givens in the previous sections. The link to Dreyer is obvious; at this fever pitch, no face in *Häxan* performs exactly the same formal task or conveys the same affective sense for the viewer. Thus, Christensen's witch, like Dreyer's Jeanne d'Arc, resists the enfolding contours of a single "type" in sequence with other equally *composed* faces. It is in this way that Christensen has made the witch *his*. Forgetting for a moment the sympathetic logic that makes this so, it is not entirely clear if Christensen recognizes that the polarity of this expressive, *metoposcopic* force can flow in more than one direction.

The Young Maiden can now name the witch. We see it on her face as she calls out to others in the house. They rush in to hear her frenzied discovery. Christensen moves the action along through a quick succession of medium shots of the commotion and extreme close-ups of the women arriving one by one at the same conclusion. He returns to close-ups of Maria wildly shoveling food into her mouth to amplify the sense of what is to come.[13] There is only one intertitle dialogue in this sequence, as the Young Maiden warns her sister that Maria "has evil eyes." The women pantomime their troubled reasoning so clearly that no explanation is needed. The viewers are meant to draw the same conclusion. Maria is indeed the witch of whom the healer spoke. We know that she is *the one*. The Young Maiden prepares to get help. Her mother, whose previous close-up revealed her to be more doubting, surveys the scene with the same look, again shown in close-up. It is unclear if she is suspicious of Maria or of the panic of her daughters. She leans out the door to look at Maria; her face is stone, furrowed and troubled. Another mask.

With the opening scenes of the film's third chapter, *Häxan* has established one line of accusation that was quite common in the early modern period. As nearly all historians of the so-called witch craze agree, the desperate search for the cause of what was otherwise an unexplained illness or misfortune was frequently the catalyst for specific witchcraft accusations between friends, acquaintances, and often between family members themselves. This is a very old and well-documented story of the fear of *maleficium* expressed through violence and killing (as much as forming an alliance or buying her off). What has changed by the sixteenth century is that this violence now bore the sanction of both secular and religious institutions. Thus, the mere fact of misfortune does not account for the specific mode by which this often-deadly reaction would take place. After all, human beings have been suffering misfortune, illness, and death long before the power of the witch was felt during this time. However, now peasant complaints of *maleficium* were by being heard by experts, who had determined, well in advance, that a satanic conspiracy was underway within Christendom.[14]

In effect the Church subsumes the role of the *un-witcher*—a role that someone such as Peter Vitter would have taken on in previous centuries. No longer the personalized, supernatural contest between *witch* and *un-witcher*, it is now the Church that, in opposition to its previous position exemplified by the *Canon Episcopi*, performs the task of engaging the witch. As *Häxan* powerfully demonstrates in the pivotal scenes that follow, this engagement is no longer characterized by the deployment of techniques to counteract individual *maleficium*, but rather becomes more fully directed toward establishing the witch as a key player in a transcendental drama aimed at destroying Christianity itself.

Witch as Vector

Häxan leaves Jesper the Printer's home where (the witch) Maria remains. The scene is now the interior of an ornate gothic church. A young priest, Brother John, is shown as he approaches a shrine of the Virgin Mary in the foreground. He begins to pray. The Young Maiden bursts into the church, distracting the priest from his prayer; he turns, startled. Although Brother John's reaction is not as tightly framed in close-up as the previous shots, the sequence rhymes visually with the Young Maiden's own shocked reaction earlier. The Young Maiden calls out to the priest and rushes over to him; he slowly turns away, reticent.

Again, the face dominates the screen in tight close-ups. The Young Maiden tearfully recounts her story to the young priest. The face of the priest is fixed and without emotion. He slowly turns and looks at the Young Maiden. He reports (via the title card) that, as the youngest inquisitor, he cannot speak to strange women. His fear of impropriety (and its perception) is confirmed as the film cuts to an older priest, revealed in the credits as "Johannes," descending the stairs in the background and observing the encounter with narrowed, suspicious eyes. The Young Maiden and the fledgling priest are now positioned in the foreground of the frame, with the small figure of the distant eavesdropping friar on the stairs visible in the space between them. The Young Maiden, not taking "no" for an answer, darts forward and grabs the priest's arm, continuing to plead her case. Stunned, the young man slowly looks down toward his touched arm, the Young Maiden's hand grasping his bare flesh under the sleeve of his cassock. This is too much for the spying senior priest and he barks loudly at the couple, causing his younger colleague to reflexively jerk back and spin away from the Young Maiden. They are caught—not only in an act of impropriety, but also by the creeping power of the witch to pervert and pollute.

It is no accident that Christensen introduces Brother John by showing him reverently approaching a shrine devoted to the Holy Virgin. Mary was elevated by Catholic demonologists such as Institoris, who belonged to a Dominican order known for aggressively promoting the cult of the Virgin, to the status of the "perfect" woman. The Holy Virgin therefore provided a stark contrast to the lustful, credulous nature of common women who were often associated with the temptations of Eve, indexed in this discourse as the first to succumb to Satan's subterfuge.[15] Brother John reacts in this scene in *Häxan* as if the Young Maiden were handing him forbidden fruit, reflecting the ferocious sexual repression demanded of the priest. Not long before, it would not have been uncommon for priests to engage in forms of clerical concubinage. Although such practices fueled heretical movements such as the Waldensians and the Brethren of the Free Spirit, they would not have necessarily drawn scorn from the general community.[16] By the time of the Reformation and Counter-Reformation, clerical attitudes regarding sexual desire had hardened to the point of elevating the status of sexual neurosis to a virtue. Clearly, Brother John's repressed distain for the beautiful Young Maiden will make a surging, deadly return later in the film. While these scenes are certainly put in the service of Christensen's unfolding thesis, which directly connects witchcraft to clinical hysteria, they also ring true in their dramatization of primary evidence from the period.

The Young Maiden rushes toward the older friar—she will make someone listen. The force of the Young Maiden's accusation is again conveyed through her face rather than title cards. The relative status and position of each character is emphasized by the camera's dramatic angles. The Young Maiden's close-ups are shot from above as she gazes up at Johannes, beseeching him to take action. In turn, the inquisitor benignly, almost dismissively, addresses the girl from his elevated position. A tight smile crosses his lips as she relays the accusation. Recalling similar scenes of Father Henrik's first confrontation with Satan and anticipating the charged confrontation between Maria and her inquisitors, Christensen consciously uses this technique to index status throughout *Häxan*.

The Young Maiden is ushered into the private workspace of the inquisitors. The mundane work of the church continues as she enters. Christensen shows this by lingering somewhat peculiarly on a friar having his head shaved to form his tonsure. Father Henrik, recognizable as the priest who was harassed by Satan himself in the previous chapter, rises to receive them. The scene returns to Brother John in the main hall, now with a dumbfounded

The Young Maiden looking up in *Häxan*, film still (Svensk Filmindustri, 1922).

Johannes looking down in *Häxan*, film still (Svensk Filmindustri, 1922).

expression on his face. He carefully regards the arm that the Young Maiden had only moments ago grasped in her excitement. "How wonderful!" he exclaims. "It felt like fire when the young maiden took my arm." His eyes widen, a look of passion building on his face—he whirls around toward the room upstairs. Meanwhile, the two older inquisitors now remain with the Young Maiden. Father Henrik is seated; he is obviously the senior member of the inquisitorial staff and it is he who will do the talking. He warns the Young Maiden of the seriousness of her accusation and demands that she swear by the cross that she and Maria "are not deadly enemies." Before we have her answer, the film abruptly cuts to Maria, still eating in Jesper's kitchen.

The scenes of the Young Maiden in the church are pivotal in *Häxan* as they reinforce several crucial elements of Christensen's overall thesis. With an economy of shots, the conflicted status of women is directly introduced into the film, particularly through sternly patrilineal visual motifs evident in the shots of the Young Maiden interacting with the older inquisitors. The mixture of condescension and bemusement exhibited by these men works to visually intensify the Young Maiden as a frail and hysterical woman. The priests take her seriously in the end, as they must take all witchcraft accusations seriously. Importantly, while Christensen makes it clear that the priests are critical toward the girl, he is also careful to show the audience that they are, within the realm of their own assumptions about the world, dedicated to investigating and verifying her claims. Contrary to some prevailing historical claims being made in the period contemporary to the film, the inquisitors are not shown to be gullible, fanatical, or overtly misogynistic in the hearing of the Young Maiden's accusation.[17] *Häxan*'s witch hunters act in accordance with their own procedures for investigating truth and falsehood and not simply out of malice, fear, or stupidity. While this claim must be taken in light of Christensen's at times explicit condescension, it is clear that the senior inquisitors display a complex and careful attitude toward the Young Maiden and receive her accusation in light of this complexity.

The Young Maiden's initial encounter with the inquisitors is dominated by two issues, one that is obvious with the benefit of hindsight, and another which is not directly dealt with in the film but is important to note. First, the manner in which the friars simultaneously display suspicion toward the girl's claim and a marked desire to believe what she is alleging is conspicuous when the Young Maiden names the witch to the inquisitors. It is important to distinguish between the *desire* to believe and simple belief. They are not the same in *Häxan*, just as they were not the same during the witch craze. Despite the *unbelievable* scope of demonic power, the inquisitors *must* believe that what the Young Maiden is reporting is possible.

Second, while the relative ease by which the Young Maiden is able to make her deadly accusation is presented economically, Christensen skips over some important changes in the legal systems in Europe at the time. Well into the 1400s, proffering a formal indictment against another individual required the plaintiff to submit to an accusatory form of criminal procedure. Derived from Roman law, this procedure presumed such offenses as matters between the accused and accuser. Thus, the idea that "crime" was a matter between society and the accused did not exist, making the present-day distinction between criminal and civil complaints meaningless. Raising a formal complaint required the accuser to furnish proof of the allegation and, importantly, to submit to severe penalties agreed in advance if the judge was unconvinced of the complaint's merit. In short, it was complex, expensive, and very risky to enter into this formal framework in order to address disputes or everyday injustices. As a result, most ordinary people did not do so, choosing instead to pursue local and less formal modes of redress.[18]

With the emergence of the witch in the fifteenth and sixteenth centuries in Europe, this procedure changed dramatically. As the witch was understood as an agent of Satan, the crime of witchcraft came to be understood as a crime against Christendom, and thus society itself.[19] Logically, the procedures for discovery and eradication of such rebellion shifted, with an emphasis on the totality of the offense and the responsibility of civil and religious authorities to protect pious Christians from a power that, without the establishment's intervention, would overwhelm the faithful regardless of their individual acts, intentions, or beliefs. This is the logical basis for the hallmark procedures of the inquisitions, as depicted in *Häxan*. Further, the responsibility of the accuser shifted from that of providing hard proof to the expectation of merely reporting what they suspected to the proper authorities. The inquisitors, firmly positioned as experts for the first time, would take this suspicion forward administratively. In an impersonal sociological sense, this is an example of Weberian rationalization. At the level of individual relations and passions, it meant that the inchoate suspicions of the fearful, the resentful, or the spiteful could now be duly received without the formal restraint of a judgment recoiling back on the one giving voice to the charge. Clearly, to suspect someone of *maleficium* was not new. Rather, the novelty that contributed to the explosion of witch accusations was that authorities were now eager to act on suspicions independently. Thus, according to its own subterranean expression, *Häxan* indicates this significantly transformed context of the complaint in the scene of the Young Maiden formally incriminating Maria the Weaver.[20]

In contrast to much that is only gestured toward in the Young Maiden's initial encounter with the inquisitors, Christensen telegraphs the complexity of Brother John's encounter with the girl much more explicitly. He acts properly toward the Young Maiden when he initially attempts to ignore her, and yet it takes only the slightest touch to transfix and overwhelm the young man. His elation in the aftermath of the encounter is clearly meant to index Brother John as ignorant. It is also a not-so-subtle bit of foreshadowing on Christensen's part, as the grammar of the film would not allow such quick, decisive passion to simply dissipate. In this sense, Brother John's passion mirrors the force of the Young Maiden's own feelings in the face of the "witch" Maria, and Christensen skillfully links the affects of each character in the visual expression of their encounter.

The meeting between these two young people strongly echoes the previously introduced theme of sexuality. A straightforward reading of this scene implies that witch accusations were often generated by unresolved sexual desires and that such passion would be redirected in this pathological, perverse manner.[21] Although historians sharply debate this as a causal factor in witch accusations, it is clearly a position that *Häxan* favors in light of the scenes that follow. The reintroduction of sexuality in this way engages the issue from another angle; the film emphasizes the crucial role sex played in discerning what constituted witchcraft and its status as a knowable category of (*malefic*) human practice. As Walter Stephens has argued, it was crucial for inquisitors to consider sex in the context of witch trials as sexual desire. The actual act of sexual intercourse was one of the primary empirical means by which witchcraft could be made visible.[22] *Häxan* develops this theme and refers back to the fateful encounter between Brother John and the Young Maiden as the film progresses.

Laying Hands on the Witch

As the third chapter of the film draws to a close, we are witness to Maria's arrest under the charge of witchcraft. Although not shown, it is apparent that the inquisitors have found the Young Maiden's accusation credible and they have sent men to apprehend Maria while she is in the family's kitchen, eating what appears to be the largest bowl of soup in recorded history. The magistrates sneak in and roughly bundle Maria into a large sack; still sitting at the table gorging herself, she did not notice the men approaching from behind. The mother enters dramatically from the bedroom; first shaking her fist at the struggling old woman and then spreading her arms wide in a Christlike gesture, she warns the men to bind Maria lest her feet touch the floor and her demonic power returns, allowing the old woman to turn them all "into mice."

Her warning reflects a long-standing worry among Church and secular authorities that they would themselves become bewitched due to the nature of their work. As Clark has demonstrated, a great deal of demonological thinking was devoted to justifying the fact that civil and Church officials, despite their fears, by and large were not bewitched. In the moment of apprehension, the magistrates were immune as they were instruments of the sovereign, understood as an extension of the Divine. As such, a witch's power was drained at the moment of "touch" by the instruments of the Divine.[23] Being "but" an ordinary woman, there is no way the mother would be aware of the existence of this immunity.

Maria is brutally launched into a small cart that has been pulled into the kitchen. The intensity of the scene is conveyed to the audience, particularly as Christensen returns again to the strategy of paralleling the action with a series of quick, tight close-up shots of the faces of the women. The sister, crying out for vengeance, is shown dramatically double-framed by the boundaries of the shot and her Hapsburg-style headdress. The Young Maiden, who has presumably returned to the house with the magistrates, excitedly witnesses the scene, a look of blood lust palpably dominates her face, reinforced by her furtive gnawing of her fingernails. Finally, the stony face of the mother appears; she pumps her fists to urge on the men. A title card relays the mother cursing Maria as a "damned mistress of the Devil." We see a close-up of Maria's withered face. She is pleading.

Christensen is, through his actors, following Satan's idiom. This is very ambiguous, as it is entirely unclear if one can "imitate" Satan in any sense without in some direct way simply being Satanic. The ambiguity is mirrored in *Häxan*'s tortured relation to "*the* truth." The question of empirical certainty and reenactment haunts the status of the film as evidence. Although he overtly disavows this association, Christensen has to a significant degree in *Häxan* sided with the inquisitors, figures who themselves were caught by the witch. He is relying on the fact that the truth of the witch will take its most visible form by acting her out mimetically. Such performance revealed the truth for inquisitors, who by this period did not require evidence of specific *malefic* acts to make a conviction of witchcraft. It was also sufficient for Flaherty, who knew that the visceral force of *Nanook of the North* depended on the felicity of his Inuit interlocutors reenacting themselves. In each case, acting the ideal type breathes life into the emptied, clichéd figure.

It is a tired truism to claim that cinema directors "act like God" in creating their films. However, displaying an intuitive sense of his craft and his subject, Christensen chooses to "play" Satan instead, using his idiom to breathe life into his witch. The interrogation of Maria in the next chapter, the old woman confessing through Christensen's possessive ventriloquism, dispels what little doubt remains regarding the truth of the witch in *Häxan*.

Demonology

> Trace and aura. The trace is the appearance of a nearness, however far
> removed the thing that left it behind may be. The aura is the appearance
> of a distance, however close the thing that calls it forth. In the trace, we
> gain possession of the thing; in the aura, it takes possession of us.
>
> —WALTER BENJAMIN, "The Flâneur"

The fourth chapter of *Häxan* gets to the heart of the matter. It is here where
Christensen's skill as a filmmaker, his facility with historical artifacts and
documents, and his understanding of witchcraft and the burgeoning fields
of neurology, psychology, and anthropology come together in the full articulation of the film's thesis. The director brings *to life* the rich visual culture of witchcraft. He draws links to the spectacular illustrative diagnostics
of Duchenne and Charcot and his visual interpretation of Freud's theories
of neurosis and the human psyche, which again and again find their way into
the film. Playing with an ambiguity inherent to deriving truth from testimony (as ambiguous in Freud and Malinowski as in the transcripts of witch
trials), Christensen takes the invisible objects of such testimony and flashes
them on to the screen. Taking Freud's talking cure a step further, Christensen
seems to be offering a cure for the secularized Christian *blindness* at the heart
of positivist human science. Just as the inquisitors are enchanted with the
witch, and by her evil, in the same way his audience is pulled in, not as

misguided inquisitors, but through a lens of science, the details of which we have yet to see.

The traces of demonological thinking and the diagnostics of nervous disease are by now discernibly etched upon the surface of the film through Christensen's complex method of inscribing such traces in the composition of images in *Häxan*. Suggesting that these etchings are "finding their way" to the surface is generally correct, but our meaning is specific. It is not that secrets rise to the surface from the murky depths beneath the image. Christensen plumbs these depths for the viewer from the beginning, placing his labor in clear view. By working with figurative givens of witches and demons, the director formulates his visual thesis. Like Freud's memory traces, the "visual etchings" here are conceptualized as etchings into a material, an act that equally implicates the eye and the hand in the apprehension of the tactile, haptic images Christensen produces in *Häxan*.[1] The director is not revealing secrets in *Häxan*; he is carving outlines into the image of figures that have been hiding in plain sight all along, longing not to be seen but rather interpreted. This allows for the power of a figure such as the witch to be perceived, perhaps quite suddenly, in a manner that loosely corresponds to the second term in Walter Benjamin's formulation above: aura. This aura, this sense, distinguishes *Häxan* from other attempts to empirically document the witch, following less the logic of dogmatic objectivity than the Freudian understanding of memory that posits it as a trace that is etched someplace in the unconscious and yet simultaneously saturates that very unconscious with its substance. Where most factual accounts would seek to bury any sense of aura, *Häxan* blasts it into the open as an uncanny, mobile power.

The fourth chapter of the film commences with Maria being "processed" for interrogation by the magistrates. The procedure entails a careful search of her body for "witch powder," carried out by two "honest matrons" who strip the old woman and comb through her hair. The search occurs off-screen, implying that the examination will be much more invasive than can be shown, heightening its perverse thrill. Christensen correctly shows that it was common in sixteenth-century Germany for witches to be completely stripped and searched, not only for "witch powder" or other *malefic* instruments, but also for the telltale mark that Satan was believed to etch on the bodies of his followers. Areas of thick hair growth on the body were of particular interest, and the head, armpits, and pubic region of women arrested under the suspicion of witchcraft were often shaved in the course of the search.[2] Maria is at least spared the indignity of having her head shaved.

"The suspect's nights are now dictated by the inquisition judges," the title card informs us. *Häxan* then moves to a series of images depicting Maria's interrogation. The questioning takes place under torture from the beginning—our first look at the now imprisoned Maria shows her confined to the stocks. Two "honorable men" attempt to elicit a confession from Maria using a vicious pantomime of a "good cop/bad cop" routine. The violence of the men jerking the frail old woman back and forth is emphasized by Christensen's rough match shots, each jerking Maria into frame, bouncing her back and forth between them. Maria appears confused and defeated throughout this sequence.

The next title card details the procedure: "If she stubbornly denies her charges, they will use a kind of mental torture." In fact, several "sacramental" items are then shown. The old woman would certainly be affected by the objects, not because of their innate power but because of the power of divinity Maria would believe they held. Although often overstated as an influence, Christensen's reliance on the *Malleus Maleficarum* as a source is quite obvious here, as sacramental objects such as the ones depicted are discussed at length.[3] While not particularly concerned with the details of the witch that *Häxan* is gearing up to display through Maria's confession, the *Malleus* offers a great deal of practical, albeit at times dangerously contradictory information (such as asserting a great power to the ritual utterances in sacramental rituals while also asserting that God confers no inherent power upon words), as to how witchcraft can be practically counteracted.

The focus is not particular to the *Malleus*, as other famous demonological works such as Nider's *Formicarius* also claim an efficacy for sacraments and spend a great deal of effort detailing the proper use of such ritual instruments. It was common for theologians to distinguish between the sacraments and powerful (but lesser) rituals that were nevertheless rooted in the same logic and drew their power from the same divine source. Unlike the sacraments, however, sacramentals such as the ones depicted in *Häxan* could be used by laymen and the clergy alike and possessed an unlimited power of iteration depending on the situation at hand. Importantly, sacramentals directly counteracted the works of the Devil—they were acts *against* Satan rather than the works *by* God.[4]

To imply, as *Häxan* does, that the use of such instruments was universal or uncontroversial stretches the historical record, however, as powerful arguments were made (particularly by German Protestant intellectuals) that the belief in such "quasi-sacraments" was itself a form of diabolic idolatry and should therefore not be used to combat the power of the witch. Indeed, when compared with the earlier scenes of Karna's use of practical magic, *Häxan* itself demonstrates the inevitable conclusion that sacramental rituals such as the ones deployed by the inquisitors were ultimately rooted in the logic of *magic*, plain and simple. For the leading intellectuals of the Reformation, these techniques were superstitious at best and evidence of the idolatry of the Catholic Church promoted through its clergy. Such rituals were formally unnecessary for these Protestants in counteracting the power of a witch under judicial custody, as the inviolability of a sovereign, cascading down the ranks of his judicial machinery, stripped the witch of her ability to act by diabolic means.[5]

Alternatively, a number of prominent texts, including the influential *Canon Episcopi*, held that the powers of the witch were really nothing but deceptions created by Satan, illusions that dissipated in the face of divine justice embodied by magistrates and inquisitors. Again, in this context, sacramentals such as holy words written on consecrated parchment and consecrated wax "as Corpus Christi" were commonly used, but were not a universal practice when interrogating witches.

Scripts and Pacts

From these relatively minor scenes, *Häxan* moves to the centerpiece of the chapter: Maria's interrogation by her inquisitors. In keeping with the previous view on counter-magic and sacramentals, Maria is brought into the torture chamber backward in order to neutralize any residual *malefic* power. Christensen frames the scene with a beautifully composed tableau of the torture chamber, the instruments of torture balanced across the image. Emphasizing the claustrophobia of the scene, there is a frame-within-a-frame, the stone building supports and low, vaulted ceiling serving as a thick, black border (just slightly off-center) for the shot. The inquisitors are moving at the center of this limited line of sight, seated in preparation for the formal questioning of Maria. Similar to Dreyer's films of the same period, Christensen is deliberate when using a depth of field, highlighting the solid shapes of doors and arches without creating the sense that this depth signifies an open or free space.[6] Rather, his strategy is consistent with the static plane of the tableau.

Häxan cuts to a medium shot of the four inquisitors; they are the same men who were present when the Young Maiden made her accusation in the church. As in the long shot, they are composed in a recognizably classic sense in the frame, visually recalling the style of the woodcuts and drawings that appear in the film's first chapter. Father Henrik, the senior inquisitor, is seated behind the others on a simple version of a bishop's chair; he demands to know if Maria is now willing to confess.

Maria is silent, mustering as defiant a look as her frail, withered face will allow. The inquisitor is obviously unhappy with her attitude, dramatically rising and ordering the torture to begin. Christensen shows this first in the medium shot displaying all four inquisitors and then, after a title card, in the original long shot, the senior inquisitor gestures to the executioner who has entered the frame.

Häxan cuts to a close-up of Maria's feet as she is being secured in the stocks. Her face slowly registers pain, again rendered in a close-up. The executioner takes a particularly evil-looking pair of tongs off the wall where various torture instruments are hanging—the inquisitors silently watch. This medium shot, utilizing the earlier framing, forcefully expresses the tableau quality Christensen is seeking. The four men are completely still, recalling E. H. Gombrich's description of Vermeer's paintings: "still lifes with human beings."[7] The depth of field and static, composed subjects stabilize the sequence. The viewer is returned to a close-up of Maria, tears welling in her eyes. She asks how can she confess to something that is not true; her question (intertitle) is met with stony silence by the senior inquisitor seated completely still in his bishop's chair. Christensen inserts several match shots between the inquisitor and Maria, emphasizing the suffocating, confrontational nature of the scene. The old woman's pain escalates; she grimaces and throws her head back. The inquisitor leans forward, increasingly aggressive in his posture. The totality of this violence is evident in the

Maria's inquisitors in *Häxan*, film still (Svensk Filmindustri, 1922).

composed alterity of the scene's stylistic correspondence between accused and inquisitor.

"Let now the evil witch's body sting," says Father Henrik to Erasmus the executioner.[8] A quick reaction shot of Father Henrik follows, then the film immediately cuts to a dramatically tight close-up, edged by the black frame of a partially closed iris, of Brother John's troubled face. Eyes closed, it is unclear if he is reacting to the spectacle of Maria's pain or the crude double meaning of his superior's remark. There is a rhyming close-up shot of Maria's face, in the same position and with the same expression as the young inquisitor's. Christensen's visual link between the accused and the accuser is explicit here. The director's disdain for the lead inquisitor is also explicit, as the film then moves to a medium shot of Father Henrik taking a break from baying at Maria to swill a drink. Again visually linking the senior inquisitor to the accused, the shot recalls earlier images of Maria's ill-mannered, wild eating in Jesper's kitchen. By the final cut the viewer is brought full circle to the initial confrontation between Maria and her inquisitors.

Close-ups in *Häxan*, film still (Svensk Filmindustri, 1922).

The sequence has shifted the oscillating rhythm between tableau and face that Christensen has established with his method. Dramatic space now re-quires reading movement from face to face, often without the stabilizing term of the tableau. In this scene, *Häxan* most resembles the "flowing close-ups" deployed by Dreyer in *The Passion of Joan of Arc*, suppressing perspective and depth of field in favor of a continuous affective movement as expressed in the face. The effect is that even the medium shots that are occasionally interspersed between tight shots of Maria and her inquisitors function as close-ups by virtue of the fact that spatial distinctions fade, leaving it entirely to the spirit or the aura of the shots to carry the narrative forward. *Häxan* is not grounded in a setting here; it is grounded in the forms of life present in the shot.[9]

The lead inquisitor then directs Erasmus to allow Maria to "catch her breath" and offers to lessen the intensity of the torture in line with the full-ness of her confession. Again, cutting between close-ups of Erasmus and Maria, Christensen continues to ratchet up the intensity of the exchange. Maria, her will now broken, begins to confess her "evil deeds." Although the timing of Maria's turn is sudden, her fear, pain, and the sense of corporeal alienation from herself are chillingly conveyed. Her reversal recalls Lyndal Roper's assertion that this sense of alterity in relation to one's own body was critical to the "success" of witchcraft confessions.[10] Able to convincingly inhabit a script not her own, Maria's confession will take her curiously eager inquisitors (and the audience) to the heart of *what witches do*, regardless of

the sliding ambiguities of fact and memory. This detail uncomfortably joins *Häxan* in a very literal way to the historical scene of a witch's interrogation.

Maria's confession is the vehicle by which Christensen's streamlined vision of the European witch comes to life, overlaid onto what are by now the destabilized visual clichés from where the film began. Focused on the Sabbat, Christensen refers again to nearly all the elements that served to constitute the witch stereotype, including the Wild Ride, the pact with the Devil solemnized through sexual intercourse, cannibalism, the cauldron as the locus of the rite, and its location at Venusberg. Although it is obvious that Christensen is selective in bringing these traces to life, the effect is powerful and largely *accurate* when judged against the surviving records of the time.[11] The power of the witch, the supernatural drama of the Sabbat, and an overwhelming demonic power is never more *real* in *Häxan* than during this visual retelling of Maria's confession.

The "guilt" or "innocence" of Maria in a positivist framework is largely beside the point by this time. Indeed, it is the task of the accused to bring the inquisitors within range of the power of the witch, just as it is the mission that Christensen has set for himself and his unlikely "star" actress[12] to do the same for the audience. The safe haven of objective distance is closed off for inquisitor and viewer alike; clearly it is a distance that both desire fervently, albeit unconsciously, to close. *Häxan* covers this range through precisely the same means as would have taken place during an interrogation of a witch—through the biographical narrative of an accused woman. What these narratives enable for the inquisitor and the filmmaker are slightly

different, however, as the judicial machinery of a witch trial required evidence of criminal acts that by definition could not be witnessed.[13]

Witness to Things Unseen

As *Häxan* demonstrates in bringing Maria's confession to life, the power of cinema to witness *exceeds* that of the witch hunter. The viewer not only witnesses Maria's act of testifying but also witnesses the acts testified to for themselves (already a violation of what is true for dogmatic positivists). In actuality, what *Häxan* does is "worse" than rigging the truth in that it aligns itself not with a concept of truth or the real but with the power of the witch outside such judgments. Bearing more than a passing resemblance to the emotionally charged, unequal relationship between the analyst and the patient, the complex relation between the accused and the inquisitor is replicated in the relation between the film and the audience.[14] As is often true in the human sciences, *Häxan* works through instruments of knowing rooted in the dynamics of the confession. To this day such instruments attain the status of empirical fact, challenging the conditions by which we can assert that something is real. In light of the position *Häxan* starts from, Christensen appears to do this despite himself.

Maria begins her confession with an account of sex. This starting point is no accident. We are taken back to the recesses of Karna's dark lair, this time the image tinted with the deep blue of night. Karna is there with her assistants helping Maria through a jarringly disturbing birthing. Terrifying, unidentifiable hybrid creatures squeeze through the elderly woman's womb, dropping to the floor. For the contemporary viewer, these creatures often elicit nervous laughs. In an age of computer-generated special effects, the costuming and props in this scene appear hopelessly amateurish. Yet Christensen has carefully chosen the manner of appearance of these demonic creatures. Perhaps clumsily rendered, the wriggling demons reflect an interesting set of variations to the witch stereotype, both ontologically and visually.[15]

Interestingly, Christensen's account of Maria giving birth to demon children appears to run counter to nearly every demonological theory on the subject. Drawing heavily from Thomas Aquinas and his theorization of the "virtual bodies" of angels in the *Summa theologiae* (asserting that "angels do not need bodies for their sake but for ours"), Rémy and others argue that demonic intercourse was always sterile.[16] Often referencing the same passages in Aquinas, Francesco Maria Guazzo in his *Compendium maleficarum* acknowledges that such unnatural couplings could produce children, but that the bodily essence of devils would rule out the possibility that these children would themselves be demons.[17] The authors of the *Malleus* agree, claiming that devils themselves are unable to reproduce but can, in the form of succubi, steal semen from men and in turn use that semen to impregnate women as an incubus.[18]

Less concerned with grotesque offspring, the *Malleus* takes up this question as a way of illustrating Satan's attack on the sanctity of marriage and his potential for disrupting and destroying this holy bond. The susceptibility of wives to the Devil's erotic charms and of husbands to witchcraft that specifically hindered the performance of his duties, particularly by "stealing" his penis, are given particular attention by Institoris and Sprenger.[19] Among the major writings on witchcraft and demons of the period, only Sylvester Prierias's *De strigimagarum daemonumque mirandis* (1521) agrees with Christensen's position on Satan's ability to impregnate human women and produce offspring such as those depicted in this scene in *Häxan*.[20] It is unlikely, however, that Christensen was relying on Prierias as a source, as the sixteenth-century author was primarily concerned with Satan's ability to manipulate and pervert *language* (including erotic language) and there is no evidence that the director directly consulted this text.

As with nearly all such instances of artistic license in *Häxan*, some remarkable truth is embedded within the scene. Echoing none other than the esteemed pioneer of ethnographic filmmaking Robert Flaherty ("Sometimes you have to lie. One often has to distort a thing to catch its true spirit"),[21] Christensen opts to draw the viewer's attention to the fact that fears of the witch often focused on the threat she posed to reproductive processes. The evil of the Devil was practically limitless; the witch, by contrast, directed her attention to very specific domains of life: marriage, children, farming, and health. In an insecure age, it is unsurprising that the danger most often attributed to the witch was a threat to *fertility*.[22] Facing the prospect of a visually tedious cinematic explanation of such a wide-ranging fear, Christensen opts to "lie" in order to catch "the true sprit" of this persistent, and by his reckoning causal, element within the story of the witch. In the span of several seconds this free-floating anxiety is firmly locked in place via an unforgettably graphic image of an elderly woman giving birth to deformed, hybrid demon children.

Rapidly recounting her "crimes," Maria suggests a small bargain with her inquisitors: "If I am spared the pain, I will confess that Trina has smeared me with witch ointment." Then shown naked in bed, Maria displays obvious pleasure as Trina applies the salve over her leathery skin, the key lighting of the scene producing a sharp, reflective sheen. Although seemingly incidental to modern viewers, Christensen's attention to the application of the salve is consistent with that of the inquisitors he is portraying, as the use of the salve was one of the most verifiable signs of the witch. Modern historians have speculated that the salve may have been the pharmacological source for the visions and tales of the Wild Ride and the Sabbat, as recipes for these salves (reported centuries after the fact) were said to sometimes contain reference to real narcotics such as belladonna.[23] Johannes Nider's famous tale of the woman who attempted to prove that she could fly through the air to a Sabbat by anointing herself with a salve also appears to indirectly support this theory. In particular, the fact that the woman in Nider's story reported that she flew despite the eyewitnesses seeing only that she fell into an insensible trance (of course they beat her while she was under the influence just to make sure) appears to suggest the salve worked as a psychotropic. Cohn regards the issue as unsettled. While noting that authors as diverse

Albrecht Dürer, *Witch Riding Backwards on a Goat* (1500). Courtesy of British Museum, London.

as de Spina, Weyer, and Tostato also report the use of salves with similar results, Cohn's own research found that most salve recipes did not clearly refer to any known psychoactive substances, and he notes that none of the above authors states having actually witnessed such a trance; even Nider's legendary account was told to him by a mentor.[24]

Christensen appears to be well aware of the ambiguity of what the salve actually does, offering no definitive comment on it. Rather, the visualization of Maria's incrimination of Trina in *Häxan* is immediately followed by a sharp cut to the outcome, imagined or otherwise, of the application of the salve: the Wild Ride to the Sabbat.

High Flying

What follows in the film is an extended view of the Wild Ride using the special effect of superimposition to show women from all over the area borne up in the air, propelled toward "Brocken." The speed and fluidity of the

women's flight in this extended sequence, composed of a series of swift tracking shots of stationary actors giving the effect of high-velocity flight past a
stable frame, is as impressive over ninety years later as it must have been
upon *Häxan*'s first release.[25]

Several motifs associated with the witch that Christensen has not been
able to deploy thus far are now on full display, particularly the wild, flowing
hair of the women (a common visual metaphor for sexual promiscuity and
disorder) and the various "conveyances" (mostly brooms, but benches, stools,
and cooking forks also appear to be represented) "known" to be used by
witches in flight. It is also here where the overarching element of the witch
stereotype is fully displayed: the massed, coordinated, *female* nature of witchcraft. Wild riding women appear from all points in the frame, multiplying
in midair. Christensen's Wild Ride also references earlier source myths of
the Furious Horde, the ancient Roman creature the *strix*, and the association of night riding with the Roman goddess Diana, who possessed a revived
cult status in the late Middle Ages.[26]

It is important to emphasize that the camera itself does not appear to move
in this fast-paced sequence. Even in this scene devoted to the wild abandon
of the witch's flight, Christensen maintains the stability of his image by min-
imizing depth cues and eschewing techniques that would allow the energy
of the event to rupture the image. The shots maintain this flat, tableau perspective even when demons are visible in the foreground, their movements
minimized to balance the swift movements taking place overhead. The camera seldom moves in *Häxan*, and when it does it completely avoids tracking
movements in or out of the image that would disrupt this naturalist unity.
Demonstrating Christensen's mastery over his chosen medium, this strategy
here stands in contrast to his strategic use of such tracking shots in his earlier
film *Blind Justice*. It is clear that the director knew precisely when to move the
camera and when to keep it still.

The spectacle of *Häxan*'s special effects should not overshadow the critical
importance of the scene in relation to the film's thesis. Cohn notes that,
while established beliefs regarding "ladies of the night" flying did not themselves generate the witch hunts that came later, these much older beliefs
nevertheless proved to be crucial in the assertion that witches were numerous, highly organized, and under the command of a supernatural leader.
Images of cannibalistic night witches were explicitly a demonological variation

on long-standing popular conceptions. In the context of such widely held beliefs, it was not difficult for these theorists to plausibly assert that what the Romans called a *strix* or what peasants would refer to as the "ladies of the night" were actually misunderstood iterations of the witch.[27] Using the same tactic that Christensen himself takes up in *Häxan*, demonologists were able to plausibly join what were until then disparate things, giving force to rationalist, Universalist frameworks, first regarding the existence of the witch, then later vis-à-vis the reality of the hysteric. As *Häxan*'s Wild Ride persuasively (albeit inadvertently) shows, both ideal types relied on the same affective power for their force.

Alternating between glimpses of the frenzied gathering in long shot and closer looks at the participants in the orgiastic rite, *Häxan* moves to an overview of the Sabbat itself. The scene strongly conveys the energy, tension, and irresistibly horrible eroticism of the rite. The Wild Ride has already ushered *Häxan*'s audience into this affective zone, matching the disordered riding with the perceived sexual inversion and moral disarray such night flights have long symbolized.[28] Interspersing shots of demons wildly playing brass instruments and various drums around steaming cauldrons, and medium shots of young, beautiful women lasciviously dancing and fondling demons, *Häxan* is clearly referring not only to voyeuristic and sensationalist accounts of the Sabbat but also to the inverted time of the carnivals that were very popular in the decades immediately before the outbreak of the witch craze.[29] The wild dancing is the key, as it signals the disruption of the everyday, the inversion of what life "normally" had to offer.[30] The effect is less one of horror than of desire and abandon.

Although this is easily forgotten in the excitement of the images, Maria is still narrating this story to her inquisitors (and to the audience). She makes the curious claim that the "Devil's grandmother was there with all of her witchcraft." The identity of Satan's grandmother is never specified, but the visualization provides some clues to Maria's reference. As Charles Zika has demonstrated, well-known classical tales dating to antiquity were actively modified and put to use in the assertion of the reality of the witch in the early modern period.[31] One of the most common literary and iconographical schemas involved the ancient god Saturn and his "children." This association is briefly alluded to in *Häxan* in the earlier scene of Peter Vitter

After Michael Herr (or Heer), *Zauberey; or, Sabbath on the Blocksberg* (1626). Courtesy of British Museum, London.

divining the cause of the Printer's illness. In relation to Maria's claim that the "Devil's grandmother" was present at the Sabbat, this is almost certainly a reference to Circe who, unlike Saturn, was often *directly* represented as a sorcerer or witch in visual culture. Widely known through the popularization of works by Virgil, Augustine, and others, the mythological story of Circe's confrontation with Odysseus in Homer's telling of *The Odyssey* and her transformation of men into hybrid animal creatures was often, starting with the *Malleus Maleficarum*, cited as evidence of the timeless, universal nature of the evil of witchcraft.[32]

Häxan reinforces the enduring figure of the witch by showing a beautiful, partially naked woman, her back to the camera, working her magic over

Devils playing at the Sabbat in *Häxan*, film still (Svensk Filmindustri, 1922).

a boiling cauldron with a jumble of diabolical instruments. The audience is witness to skulls, bones, an hourglass, candle stands, playing cards spread out, and a large open book. These items, artfully composed within a sorcerer's magic circle, are ones that would be reflexively associated with necromancy, magicians, and witches. As we discussed in great detail in our first chapter, the specific link with the sorceress Circe is emphasized by the presence of the cards and other instruments of illusion and trickery, visual references that may well have been lost on *Häxan*'s original audience in 1922.

Maria does not linger on her sighting of the "Devil's grandmother." The reference would have been common sense to her inquisitors, and *Häxan* treats the allusion in the same way, leaving it to the audience to decipher. The narrative jumps to the Devil's miserable treatment of women "who had not accomplished enough evil deeds." A somewhat elliptical shot of a group of demons harassing a woman in formal Burgundian dress follows, and the violence Maria claims was visited upon such women is implied more than shown in this brief scene.

A Liturgy of the Counter-Sacrament

Arranged in the manner of a priest reciting mass, Christensen (as Satan) is conducting the rites of the Sabbat. He is positioned in the center of the frame; over a cauldron he holds a newborn child in his right hand. A large liturgical book is open, positioning the viewer as participant among the circle of lesser demons and witches who are gathered around this altar. In keeping with the logic of inversion that dominates demonological thinking regarding the procedural efficacy of satanic rites, the scene is indexed as an "anti-mass," through the shocking image of Satan sacrificing the infant.[33] Christensen is rigorously empirical here, *visualizing* powerful literary descriptions of the Sabbat and emphasizing elements that highlight the demonic strategy of making profane use of powerful elements of Catholic liturgy, ritual, and practice. In this scene, the sacraments of baptism and communion are strongly referenced, although the flesh and blood in play here is not the divine substance of Christ but rather that of a newborn, presumably unbaptized, child.

Although it is plausible to see *Häxan*'s depiction of the rites of the Sabbat as a "parody" of the Catholic mass, this interpretation does not do justice to the meticulous care Christensen takes in the composition of this scene. Unlike nearly all films since which have attempted to depict satanic rites of some kind, the director is *restrained* in what the scene could potentially contain. While the logic of inversion would seem to suggest that a satanic mass would by definition constitute a simple burlesque of the "actual" mass, it is clear that the power of such rites was not rooted in parody but in its constituting a *counter-sacrament*, a rite potentially superior to that of the Church. Demonologists such as Alphonsus de Spina in the *Fortalitium fidei*[34] carefully outline the outlandish and negating character of the Sabbat, making it clear that scandalous display might be blasphemy, but this alone would hardly threaten the true power of the sacraments.[35] The Sabbat did not aspire to merely mock Christianity (although it certainly did this), it sought to supplant and overwhelm it.[36] *Häxan* is both more internally consistent on this point than even many of the scholarly discourses (such as the *Malleus*) and careful to avoid the lampooning tone of many cinematic depictions of satanic rites in the years since, setting *Häxan* apart from later "Devil" films such as Ken Russell's *The Devils* (1971).

In its initial medium shot, the film's suggestion of an impending human sacrifice is much stronger than that of baptism. It is only when the camera moves closer to Satan conducting the rite that a clearer sense of the baptismal overtones of the scene become clear. A naked woman, on her knees before Satan, awaits the anti-sanctification of the ritual. Satan lifts the wriggling child in his hand before the assembled witches and demons (and us) and, upon completion of the vow, tosses the infant into the boiling cauldron.

Christensen avoids performing the blatant anti-Semitism that conjoins the cannibalism of the witch with the supposed ritual infanticide attributed to Jews that runs through much of the demonological literature. This durable stereotype fueled a great deal of theological speculation, particularly around the question of the efficacy of the sacraments and the potentially "jealous" misuse such rites and objects could engender.[37] *Häxan* frankly deviates from the Christian account here, as a faithful re-creation of the Sab-

Satan conducts the rite of the Sabbat in *Häxan*, film still (Svensk Filmindustri, 1922).

A counter-sacrament in *Häxan*, film still (Svensk Filmindustri, 1922).

bat based on demonological accounts, which contained almost by definition strong anti-Semitic undertones. To Christensen's credit, he does not hide behind an ahistorical ethic of "accuracy," sacrificing historical fidelity in this case to avoid the violent reproduction of such symbolizations.

Häxan does not, however, shy away from a more generalized trope of cannibalism. The image of the witch ritually killing and eating infants is one of the most resilient, mobile imaginings that circulated as an element of the popular stereotype. It is also an element that has one of the longest historical pedigrees within the assemblage, although the practice had by the sixteenth century come to be directly associated with witches. Ironically, the charge of ritual cannibalism was most persistent against early Christians. Despite scholarly refutations by younger Pliny and Tertullian in the second century, the belief that Christians ritually slaughtered infants was widespread in the Roman empire. As late as the year 160, close advisers to the Emperor Marcus Aurelius still publicly denounced Christians due to the moral threat posed by their supposed practices of cannibalism, infanticide, and incest.[38] The problem then, as in the sixteenth century, was the uncomfortably close symbolic association that could be made between the Eucharist and forms of ritual sacrifice and cannibalism that were believed to be evidence of the evils of nonbelievers. Even a careful proto-ethnographer like Tertullian noted the unnervingly literal Christian rites of the Eucharist, a literalization confirmed in the earliest account of the sacrament by St. Paul in 1 Corinthians.[39]

The image of the cannibal never dies, it only undergoes a transformation over the centuries—a slander against early Christians taken up and instrumentalized in the service of marginalizing outliers within medieval Christendom. In keeping with the doubts and desires of believers at the time, such images are perhaps the most extreme forms of unconscious identification *with* the evil that threatens in the form of the witch.[40] In addition to the fantasy of ritual slaughter of children by Jews, other groups were tarred with the label of cannibal during the late Middle Ages, including the Christian heretical sects the Cathars and the Waldensians. These groups offered substantial opposition to a wide variety of Church teachings and set the stage for both the Protestant Reformation and the viral proliferation of the witch in sixteenth-century Europe.[41] Finally, a growing literature on the "heathen" peoples of the New World consistently refers to their supposed cannibalism, often making explicit equivalences between heretics and unbelievers in Europe, and drawing further equivalence between the barbarism of the savage and the timeless universality of witchcraft.[42]

The figure that could bind together such a broad range of people under the sign of the cannibal was that of Saturn. Revived through late medieval translations of Arabic texts, the powerful Roman god (Kronos in Greek mythology) was popularly associated with tales of patricide and sexual violence. Importantly, upon hearing a prophecy that one of his sons would rise against him, Saturn was said to have devoured his own children as a way of preserving his rule. Artists at this time often exploited the figure of this infanticidal cannibal god as a visual shortcut for representing a total evil that would otherwise overwhelm the imagination.[43] Christensen is well aware of this association and the iconic representations of Saturn, composing the scene of the counter-mass in a manner that recalls classical images of the Roman god produced both at the time of the witch hunts (Evrart de Conty, *Saturn Devouring His Children*, ca. 1401; Boccaccio, *The*

Saturn Devouring His Children, from Evrart de Conty, *Le livre des échecs amoureux moralisés*, detail (ca. 1401). Courtesy of Bibliothèque nationale de France, Paris.

Castration of Saturn, 1531) and later (Goya, *Saturn Devouring One of His Sons*, ca. 1823).[44]

This dramatic sacrificial ceremony is hardly the only rite executed at the Sabbat. At the climax of this counter-baptism Maria's "voice" chimes in, informing her inquisitors (who have been visually absent from *Häxan* for some time by this point) that the "masses spat upon all that is holy." We are shown young, beautiful women recklessly stomping and spitting on a cross lying on the ground.

Returning to the cannibalism trope, Maria claims that those at the Sabbat enjoyed "a meal of toads and unchristened children" cooked by Karna. The implicated woman is then shown raising a fat, lifeless infant in her hand and tossing it into a steaming cauldron. The revelry continues with a series of close-ups showing the younger witches kissing and fornicating with the assembled demons. Satan surveys the scene from the side, a leering grin painted across his face as he pounds away at his churn. The reference to masturbation is so strong it would be inaccurate to say the scene "alludes" to anything—it is explicit. As if on cue with his ejaculatory climax, *Häxan* abruptly returns to the carefully arranged tableau of inquisitors, stonily absorbing Maria's confession. The contrast between the reckless hedonism of the Sabbat and the voyeuristic, life-denying rejection of such gratification in the Church is dramatically evident here, reminding the viewer that the force of the witch was not simply that of a simple inversion of morals but rather in the power to conjoin *eroticism* with *irony*.[45] In bringing this conjoined sense to the audience, Christensen's studied contrast here mirrors his formal cinematic strategies opposite the techniques by which inquisitors sought to produce dynamic testimony of the witch's form of life. One reflects the other in the play on the uncertainties that exist between citing formal norms, be they theological or cinematic, and self-consciously *breaking* them.

A Proliferation of Scripts

Silent for a very long time, the inquisitors begin to lead Maria through her confession, asking if she saw "how the Devil put his mark on the witches' foreheads." As numerous scholars of the witch trials have noted, the strategy of leading the accused in her testimony was common during interrogations.[46] In addition to goading, inquisitors would attempt to locate the traces Satan leaves in the form of marks on the body as a powerful piece of empirical evidence, proof that the inquisitors will go to great lengths to verify.[47] This particular procedure will return in the seventh chapter of the film, only in this case it will be used by a modern doctor looking to verify anesthesia.

Maria completely ignores the inquisitor's leading questions, instead offering the non sequitur that she saw the witches "kiss the evil one on his behind." A close-up of Maria making this statement leads directly to the faces of the men (all but Brother John, revealed in distress earlier); one by one they have a good laugh at the image conjured up by Maria's statement. Father Henrik is shown with eyes widened, a look of prurient anticipation on his face. Not one to disappoint, Christensen immediately cuts to the scene of this scandalous deed.

The director is making a judgment here as to the repressed desire and the "true" motivations of demonologists and their documented obsession with the sexuality of Satan and his witches. Christensen is not incorrect, although the reasons that the inquisitors would be so eager to hear such testimony is certainly much more complex than *Häxan*'s insinuation here that

Satan leaves his mark in *Häxan*, film still (Svensk Filmindustri, 1922).

Doctor tests for anesthesia in *Häxan*, film still (Svensk Filmindustri, 1922).

we are dealing with a group of powerful, dirty old men. While the obscene shock of such acts must have provoked a variety of immediate reactions from the authorities overseeing a witch trial, the idea that they would find it explicitly humorous seems more in line with Christensen's anticipation of his audience. If anything, the inquisitors would most likely feel a combination of revulsion and relief from Maria's statement. While confessing to having engaged in nearly unbelievable perversion would work to confirm guilt, the inquisitors at this stage still require much more from Maria. Namely, they must all *work together* to build a coherent and, by the standards of the process, consistent account of an individual's witch experiences. This phase of the interrogation would hardly be taken lightly and, with the exception of situations of panic where the volume of cases overwhelmed the experts on hand, would not have been rote or *routine*. Sixteenth-century inquisitors were therefore working along two lines simultaneously; they needed a believable confession to establish guilt and they were commanded to find out everything the witch knew about the Devil and his activities in the world.[48] Accomplishing these goals concurrently was neither easy nor an occasion for the frivolity displayed by the inquisitors in *Häxan*. Were this to occur, these experts would simply not be doing their jobs well.

Christensen brings the ritual kissing of Satan's anus to life in the blunt, non-allegorical fashion of pornography. This strategy also has its roots in the classical iconography. Satan is shown from the side and bent over, braced against a wall, having his ass kissed by a parade of women (*Häxan* does not go so far as to show the women running their tongues over his anus, which was, in the strictest sense, what demonologists believed took place). The satisfied look of the senior inquisitor is shown in a reaction shot, followed by a close-up of Maria. Her head is back and she has a look of revealing determination as she proceeds to incriminate Jesper's mother-in-law, Sissal the servant, and Elsa, "who kicked me some time ago," stating that each placed the obscene kiss on Satan's backside. As Maria condemns these women (*Häxan* makes it very clear that with the introduction of Elsa, Maria is now motivated by revenge) Christensen shows each of the alleged kisses in a close-up of Satan's ass.

Elsa's guilt is elaborated for the inquisitors with the accusation that she and her accomplice caused the premature death of Martin the Scribe by

Francesco Maria Guazzo, *Compendium maleficarum* (1626), as it appears in Chapter 1 of *Häxan*, film still (Svensk Filmindustri, 1922).

witchcraft, using a ritual that involved the sisters stealthily approaching the writer's door at night, urinating into pots, and chanting a spell while turning in place three times. This *maleficium* reaches its climax when the women violently hurl the pots filled with urine against the writer's door. Having drawn attention to their presence through the loud clattering of the pots, they run out of frame from the darkened, now urine-splattered door. As with the preceding scenes, *Häxan* does not spare the viewer the frank obscenity of this act. The viewer is directly confronted with an artfully composed shot of two women squatting over their chamber pots, their urine plainly splashed all over the writer's door. With its focus on bodily filth, abjection, and willful obscenity, Christensen's depiction is disturbing. This is neither accidental nor, within the strategy that Christensen is pursuing, gratuitous. Using well-established cinematic strategies for making elements of action intelligible without making them directly visible is *not* sufficient for Christensen.

The obscene kiss in *Häxan*, film still (Svensk Filmindustri, 1922).

Häxan must go further in evoking powerful, revolting, terrifying reactions from its audience. It is precisely through this evidence of real events that *Häxan* forces its viewers to directly confront the *power* of the witch.

Maria is now on a roll. She offers to tell her inquisitors about the witches who "yell" after her where she lives. Christensen knows, however, that the film must allow the viewer to take a breath. The chapter abruptly ends with Maria's inquisitors excitedly conferring and struggling to record the torrent of diabolic evidence flowing from the tortured old woman. In taking up Maria's perspective in this chapter, Christensen has succinctly articulated the forces in play in a witch confession that continually threatened to pull the entire enterprise apart. The unbelievable testimony of the witch consistently deranges the witness. As with scientists faced with the teeming multiplicity of nature, the inquisitors struggle to grasp fragments of this totality in order to attain some semblance of unity to delineate what is natural and what is real. They must believe in the impossible events that they are hearing, assimilating this diabolic world to knowledge while countering the relentless energy and chaos that threatens to intervene on the side of non-knowledge.

In this chapter of the film, *Häxan* explicitly crosses over from a position of observing the witch to acting as a direct vehicle for her testimony. This places the viewer in a delicate position in relation to the film, as any attempt to forge a coherent unity is stubbornly resisted by *Häxan*. To judge the film as a set of incoherent fragments, however, is incorrect, as Christensen has very rigorously established patterns that allow for the formal techniques of the film to communicate the material qualities of the witch. Christensen has staked the film on the fact that by now the viewer, like the director, wants more. Thus, *Häxan* now wagers that the viewer will not be satisfied with the safely closed unity of *explanation*, instead seeking an impossible *belief* in the reality of the witch herself. In short, the truth of the witch in *Häxan* has moved from one of *trace* to that of *aura*.

PART TWO

A Mobile Force in the Modern Age

1922

[W.H.R.] Rivers is the Rider Haggard of anthropology; I shall be the Conrad.

—BRONISLAW MALINOWSKI

Häxan was released in a year when innovation was in the air. The year 1922 saw the release of one of the first "ethnographic" films, *Nanook of the North*, as well as the refounding of anthropology by Bronislaw Malinowski. But more than the coincidence of the year 1922, *Häxan* was released at a moment when concerns with others, the supernatural, the recesses of the mind, and even evil itself were coming to be addressed by new means, with fresh commitments and a new set of anxieties about the capacities of observation and science coming to the fore. Our introduction to the second part of the film focuses on the innovations of 1922 that parallel and frame Christensen's project. But we also consider the particular claim to scientificity that *Häxan* pursues—one that does not merely parallel Malinowski's Conradian trip up the river, but attempts much the same thing. In the same way that Malinowski deprecated his seniors for what he saw as an anemic engagement with an unrecognized world, *Häxan* sought to represent the witch and evil itself, to grasp them through the trope of hysteria, to unleash and control them at the same time.[1]

1922 and Cinema

Even if *Häxan* had been able to avoid the difficulties of censorship and the limited viewership that arose due to its controversial subject matter, the film faced a great deal of competition for notice: 1922 was the year when now-classic films such as *Nosferatu* and *Phantom* (both F. W. Murnau), *Dr. Mabuse, the Gambler* (Fritz Lang), *Foolish Wives* (Erich von Stroheim), and *Cops* (Buster Keaton) were released. But much more important, 1922 was the year Robert Flaherty's *Nanook of the North* made its debut—and it is through Flaherty's *Nanook* that we begin our comparison to Christensen's *Häxan*. In *Nanook*, Flaherty produced a film within a genre that did not yet exist, exhibiting the definitive "creative treatment of actuality" that John Grierson would famously attribute to the documentary,[2] but only several years after 1922 and in direct reference to a different Flaherty film (*Moana*, 1926). Critics at the time tended to assign *Nanook* to the category of "spectacle films," "travelogues," or "scenics," and with good reason. Flaherty was hardly the first filmmaker to have produced seemingly factual films pertaining to distant and exotic places. Indeed, the travelogue as a genre was coined in 1907 by Burton Holmes, and in the decade immediately prior to *Nanook*'s release numerous films depicting the exploration of far-off and unusual places and cultures found a popular audience. Among the most influential of these early travelogues was Herbert G. Ponting's record of Captain Scott's expedition to the Antarctic in 1910–11, a film that historian Rachael Low regarded as "one of the really great achievements, if not the greatest, of British cinematography during this unhappy period."[3] Yet it would be inaccurate to claim that even Ponting's moving record of Scott's catastrophic expedition served as an influence on Flaherty. Ponting had only published his photographs from the expedition in 1921 under the title *The Great White South* and his film, *The Great White Silence*, would see release in 1924, two full years after *Nanook* and *Häxan*. Strangely, although shot a few years later, it is arguable that, in its final form, it was Flaherty who exerted an influence on Ponting rather than vice versa.

Keeping in mind the complicated chronology of many early travelogues, Paul Rotha's claim that *Nanook* occupies an originary position in regard to documentary film makes sense. Specifically, Rotha states, "[In] *Nanook*, for the first time in film history, a motion picture camera was used to do more than just record what it finds before its lens. This is the major significance of *Nanook*."[4] For all of the problems that *Nanook* raises in terms of the "truthfulness" or "facticity" of its depiction of Inuit life, Rotha's grand claim is not so far-fetched. Yet, as we have seen, the link that is often made between films such as *The Great White Silence* and *Nanook* can be quite misleading, as many films only saw the light of day or gained some audience in the years *after Nanook*'s release. Flaherty may have planted the "seed" of the documentary with *Nanook*, but it is quite clear that the conceptual soil that allowed that seed to grow was only partially composed of other works. If we want to know what made Flaherty's achievement possible we must look to these other sources, understood as constituting a milieu rather than a direct model, as essential to those that we have linked to *Häxan*'s genesis. Taken as expressions of the wider milieu of "1922," it is more than the superficial coincidence that these two otherwise wildly different films are linked.

1922 and Anthropology

Bronislaw Malinowski published *Argonauts of the Western Pacific* in 1922. This fact would make it impossible to argue that this foundational anthropological text served as a direct model or influence for *Häxan* (or for *Nanook of the North*). And yet, the shared substance of innovation that gives the period a coherence in our reading links Malinowski's groundbreaking work to these films along a number of essential points, as we detailed at the beginning of this book. We have already raised several of these points in the course of following Benjamin Christensen's thesis step-by-step in *Häxan*, with emphasis on the surprising continuity that existed between methods rooted in cultivating experience and gathering testimony deployed by theologians and inquisitors during the time of the witch craze, and the echo of these methods in the establishment of the human sciences. In this regard, the legitimacy of evidence derived from the techniques of participant observation that Malinowski outlined and deployed in the Trobriand Islands appears to bring the long struggle to verify and examine that which cannot be seen full circle—a true expertise of nonsense that took up the invisible forces of radical Others without explicit reference to Christian metaphysics. *Häxan*, with its consistent reference to science, gives voice to Christensen's antipathy toward the superstition of the Church and aspires to a position of resonating similarity to Malinowski in *Argonauts*. Yet the question how strong this tie is must be raised, considering the wide formal differences that exist between *Argonauts*, a written ethnographic work supplemented by photographs, and *Häxan*, a cinematic work supplemented by historical texts.

The logic of associating *Häxan* and *Argonauts* should become clear once we recall that Malinowski's revolutionary methodological innovation was to collapse the established division in nineteenth-century anthropology between those who obtained primary data regarding radical cultural difference and those who reflected on, analyzed, and interpreted these facts. Earlier anthropologists were quite suspicious of facts derived from personal observation and witnessing, mistrusting in nearly equal measure the competence of the outside observer to recognize fact from fiction and the native's willingness or ability to reveal the truth of himself through testimony.[5]

There are two decisive facts that deeply complicate this understanding of Malinowski's use of the image in ethnographic work. The first is the fact that the most palpably visual element of his practice was paradoxically his writing. This returns us to George Stocking's claim that Malinowski's "I-witnessing" style functions as a "narrative technology" that seeks to position the reader "imaginatively within the actual physical setting of the events."[6] Malinowski's writing is littered with examples of this cinematic style, insistently demanding that the reader imaginatively visualize the scene and, much like a camera tracking in on its subject, follow the ethnographer/director. While already well-known among anthropologists thanks to the work of Stocking, Michael Young,[7] and Christopher Pinney,[8] it is worth briefly highlighting the tone and style Malinowski uses while introducing the Trobriand Islands early in *Argonauts*:

> Imagine yourself suddenly set down surrounded by all your gear, alone on
> a tropical beach close to a native village, while the launch or dinghy which has
> brought you sails away out of sight. Since you take up your abode in the
> compound of some neighbouring white man, trader or missionary, you have
> nothing to do, but to start at once on your ethnographic work. Imagine further

that you are a beginner, without previous experience, with nothing to guide you and no one to help you. For the white man is temporarily absent, or else unable or unwilling to waste any of his time on you.[9]

Imagine yourself. Imagine *further*. It is the *imagination* that authorizes the epistemological value of *being there*. Just as the ethnographer can only discern the truth of a culture by witnessing it, the interested reader can only participate in, and evaluate the truth of, such claims by imaginatively entering the image that the skilled ethnographer conjures out of this primary experience. Importantly, the initial suspicion of observed facts made visible by ethnographic means triumphantly *evaporates* in the confident, imaginative expression Malinowski offers in *Argonauts*. Indeed, among Malinowski's many achievements, his active, quasi-cinematic writing style still stands out as one of the most striking and valued elements of his work nearly one hundred years later. It does, however, beg the question of what role photography has to play in all of this.

Malinowski's actual field photographs seldom clarify, illustrate, or announce the truth of their subjects at all. Perhaps more disturbingly, what they do accomplish is to close the gap between photographer/anthropologist and the "Other." Christopher Pinney suggests that the value of Malinowski's photographs is that they unintentionally reveal the "optical unconscious" underlying the very act of trying to grasp the world of these others. Working with native photographer Butch Hancock and taking inspiration from his friend, the Polish modernist novelist, painter, and photographer Stanislaw Witkiewicz, Pinney highlights the "fragmented," "inventive," and "disturbing" effect of Malinowski's photographs, noting that "the suturing of the theorist with the 'man on the spot' in the new figure of the fieldworking anthropologist problematized photography's role as messenger between the 'field' and the study."[10] As we have already seen, the complex issue of ascertaining the truth value of a fact visible only through the eyewitness accounts of others is one that was hardly new in 1922. What is striking is the fact that the technological ability to "directly" record events without the subjective eyewitness, at first hailed as a true innovation in the human sciences, only served to intensify the power of the invisible forces that the technology was supposed to abolish. If anything, as Pinney's claim makes clear, the nature and diversity of these forces has only grown and intensified. After all, the viewer in 1922 now also had to deal with the "unconscious" as yet another invisible element of a world whose "truth" seemed to many in the aftermath of the Great War more remote than ever.

1922 and Other Witches

Although we would argue that Malinowski went too far in casually associating Rivers with this attitude, simply ignoring, suppressing, or actively trying to debunk the excessive quality of the Other in historical and anthropological engagements of witchcraft does not pave the way to more credible, empirical, or factual accounts. There is no better example of how badly this can go than another artifact from 1922, Margaret Alice Murray's book *The Witch-Cult in Western Europe*.[11] Presented with the subtitle "A Study in Anthropology," this influential text is anything but. Taking up the "armchair antiquarian" style of Sir James Frazier in *The Golden Bough*,[12] Murray's basic argument is that a highly organized pre-Christian fertility cult was either mistaken for, or twisted into, a conspiracy of witchcraft during the witch craze in Europe. Murray's argument that everyday practical magic regarding fertility was folded into the witch stereotype by itself is not particularly objectionable, as we have explored in more detail elsewhere in this book. The author's assertion that this cult was a unified, organized, and ultimately benign phenomenon, however, is one that now flies in the face of nearly all modern historical research on the subject. What historians and researchers of witchcraft in more recent times have found outrageous about the book is summed up in this basic claim: "The objectors [to the reality of the cult] . . . overlook the fact that the believers in any given religion, when tried for their faith, exhibit a sameness in their accounts of the cult, usually with slight local differences. Had the testimony of the witches as to their beliefs varied widely, it would be *prima facie* evidence that there was no well-defined religion underlying their ritual; but the very uniformity of their confessions points to the reality of the occurrence."[13]

This, in a nutshell, is Murray's argument for the existence of a widespread, nonthreatening, victimized *religion* spread across the whole of Western Europe. Outside the claims made in confessions and during witch trials (which, even in Murray's own book, vary widely in what they actually say), Murray offers nothing else to establish the existence of a pre-Christian church shadowing Christianity. Macfarlaine,[14] Thomas,[15] and Cohn[16] have all spent a great deal of effort to demonstrate that there is simply no corroborating evidence for this claim and, while the influence of a wide range of pre-Christian fertility rites and everyday magical practices remains a hotly debated topic to this day, the idea that the witch craze suppressed a highly elaborated and organized religion is no longer widely accepted among scholars.

This is not to say that Murray's book was without influence. To the contrary, her account of an animistic fertility cult ruthlessly suppressed by Christianity has served as a key inspiration for the invention of modern witchcraft as a religion among others in the contemporary spiritual marketplace. As Tanya Luhrmann has shown, modern iterations of witchcraft, magic, and other New Age variants on the occult very often look back to Murray's victimized fertility cult as the precursor and ancestor of their own worship.[17] This is an interesting phenomenon in its own right. It must be pointed out, however, that this is *not* the witchcraft that we find in *Häxan*. While Christensen also at times engages in misstatements of historical fact, it is obvious that the main body of his account has little in common with Murray's neutered understanding of the power attributed to the witch. As with the other works discussed here, Murray's perspective on the witch is of its time, but in a much different way than Christensen or Malinowski or, to be fair, even Rivers. *The Witch-Cult in Western Europe* presents a perfectly safe, explicable, and secularized figure of the witch, victimized by oppressors for being "different." As such, it is untroubled by invisible, nonsensical forces, providing the perfect historical root for an invented identity politics where "religion" becomes primarily a concern of "the self" and what one consciously "believes."[18] Murray's witch is a transparently knowable, good object capable of serving as a personal ethical anchor in the world. By contrast, it is obvious that the Wiccan or Satanist of today can find *very little* in *Häxan* to validate this notion of "self." Rather, the bad object Christensen conjures, in our view much closer to the empirical figure of the witch as such, can offer *nothing* of this kind to the seeker.

Wrestling Dark Forces

Even in 1922, in the midst of scientific and technological advancement, the hidden, invisible forces that bedevil everyday living continued to hold a central place in the human sciences. Understood in this way, Christensen's project in *Häxan* is not as distant as it seems at first glance from the work of Malinowski or Flaherty appearing at the same time. The film to this point has worked to visually bring the witch, seemingly a figure of a past, to life. In the second half of the film, Christensen will turn to contemporary concepts of the invisible (psyche, instinct, culture) in an attempt to simultaneously delink the power of the witch from her subjectivity and offer a visualization of the nonsensical as it existed in his time. Seen in this light, it is no accident that the final three chapters of the film turn inward, focusing on possession, technique, and ultimately the interior of the mind itself in the form of hysteria. In some sense Christensen is offering a popularized Freudian interpretation of how the invisible power of the witch is actually an internalized force of the mind, but to simply leave the interpretation at that proves too pat, too "clean" in relation to what we actually witness in what remains to be viewed in *Häxan*. Christensen's attempt to picture this force remains too multiplicitous, too mobile to simply find a new home in "the mind."

Let's compare further the commitments Christensen and Malinowski shared. Malinowski explicitly understood this invisible force to be mobile, invisible, and ultimately dark. Recall the statement that serves as the epigraph of this section: "[W.H.R.] Rivers is the Rider Haggard of anthropology; I shall be the Conrad."[19] This is an extraordinary claim for a human scientist to make. Malinowski refers to *Heart of Darkness*, Joseph Conrad's classic novella detailing the violence, racism, and ultimately the weakness of "civilized" men in the face of the dark, invisible forces that await them as they penetrate further into "unknown" regions seeking riches and rule. *Heart of Darkness* has been extensively analyzed in the century since its publication, and its outline plot of the ivory trader Marlow's journey to the deep reaches of the Congo to retrieve the shadowy Mr. Kurtz, who has "succumbed" to the dark forces of the native interior, is well-known and does not require analysis here.[20] What is interesting to us is why Malinowski, the sober ethnographer seeking to demystify racial and cultural others, would so forcefully associate himself with the deeply problematic position Conrad takes on forms of contact and encounter. Speaking through his narrator, Marlow, Conrad himself stakes out a provocative claim of expertise in relation to the dangerous, invisible forces of the jungle:

> The thing to know is what he [Kurtz] belonged to, how many powers of darkness claimed him for their own. That was the reflection that made you creepy all over. It was impossible—it was not good for one either—trying to imagine. He had taken a high seat amongst the devils of the land—I mean literally. You can't understand? How could you—with solid pavement beneath your feet, surrounded by kind neighbours ready to cheer you or fall on you, stepping delicately between the butcher and the policeman, in the holy terror of scandal and gallows and lunatic asylums—how can you imagine what particular region of the first ages a man's untrammelled feet may take him into by way of solitude—utter solitude without a policeman—by way of silence—utter silence, where no warning voice of a kind neighbour can be heard whispering of public opinion. These little things make all the great difference. When they are gone you must fall back upon your own innate strength, upon your own capacity, upon faithfulness. Of course you may be too much of a fool to go wrong—too dull to even know that you are being assaulted by the powers of darkness.[21]

Published years after his death, Malinowski's famous field diaries make it abundantly clear that he struggled with the strong suspicion that, in undertaking his groundbreaking journey of what was supposed to be *scientific* encounter, he had in fact taken his own seat among the devils of the land. As with Conrad, risk was simultaneously virtue. Malinowski was certainly no fool, after all, and his "going wrong" was hailed retrospectively as a great triumph for the science of man, the experiential basis for a very special kind of expertise. But what *is* the object here? Certainly the Trobrianders themselves were not "the devils of the land" so feared in Malinowski's co-opted formulation of Conrad, just as the individual women accused of being witches were not so for Christensen. What then does Christensen or Conrad or Malinowski mean when they assert that their accounts of invisible forces are "literal"? What unites all these seemingly disparate works is the assertion of mastery over invisible, nonsensical, decisive forces in the domain of human life. The character of modern life in 1922 seemed to have raised the stakes, and advances in the sciences greatly transformed the concepts by which one could attempt a visualization of these forces, but the violent unknown, sensed yet inchoate, at the heart of things betrays a striking consistency across these varied modes of expression. It is here where *Häxan*'s own status as a work *of its own time* becomes quite clear, which Christensen consciously asserts in the latter half of the film.

Returning to the earlier quote, on the face of it Malinowski's slight aimed at the pioneering ethnographer and psychologist W.H.R. Rivers, who died in 1922 (though not likely as a result of the insult), appears to be motivated by petty professional envy felt toward an eminent scholar.[22] Indeed, Rivers's innovative approach to field research as a member of the Torres Straits Expedition in 1898 and his own subsequent studies in Melanesia and Polynesia in 1907–8[23] goes some way toward troubling the Euhemerist myth of *Argonauts* as the "ground zero" of an ethnographic field research method. Viewed alongside Rivers's innovative work in neurology and psychology, it hardly seems accurate, let alone fair, to compare the older scholar to the pulp adventure author H. Rider Haggard (1856–1925), best remembered today for his novel *King Solomon's Mines*.[24] What did Malinowski mean by this acerbic comparison?

Again, it is the nature of the object of inquiry that is at stake here; looked at from this angle, Malinowski's comment takes on a different light. Whether writing on anthropological or psychological topics, Rivers consistently disavows the invisible, mobile power of the forces at stake in the investigation. Anticipating by over two decades the arguments of British social anthropologists such as E. E. Evans Pritchard in their mid-century structural–functional heyday, Rivers argued that seemingly magical or mystical beliefs were, from the internal viewpoint, simply logical outcomes of the epistemological categories available to indigenous others and therefore not particularly different from our own in terms of their social function.[25] At their root, there was nothing dangerous or mysterious about such things, be they medical, magical, or religious in their outward expression.

Likewise, when his attention was turned toward war neurosis and psychological issues, Rivers took exception to the violent power of Freud's understanding of the dominance of the sexual drives in the human psyche, expending a great deal of effort in taming and revising Freud's drives into a theory of natural "instincts" that were neither particularly sexual nor violent in character.[26] While always a respectful dissenter, Rivers could not accept that society was founded in the wake of a violent, sexual crime, nor that the Oedipal complex that supposedly saw the founding of the social would similarly be at the dark root of individual neurosis, war-related or otherwise.[27] Indeed, working with war neurotics at Craiglockhart War Hospital near Edinburgh during the Great War, Rivers distinguished between witting repression (still careful to link such repression to instinct rather than drive) and involuntary, unwitting "suppression," making the center of his therapeutic practice the simple act of lifting the "banishment" of terrible war memories in order to see them clearly.[28] The very notion of "trauma" in the Freudian sense is quietly and calmly undermined by Rivers in preference to the simple act of "seeing" what is "really" there in the mind but suppressed for misguided social or medical reasons. Whether reading Rivers on Melanesia or on shell shock, he provides a clear, confident, almost *comforting* framework for seeing everything that is there and nothing more.

And this is precisely what Malinowski disdains in Rivers—*his placid lack of imagination.* Unfair though it may be, Malinowski has judged Rivers to completely miss essential empirical elements of the wide range of phenomena that he ably addressed through the course of his long career as a human scientist. While on the surface of things both men would agree that they are seeking to "demystify" (in literal terms) the lives of others, Malinowski seems much more troubled by the elusive, invisible elements of life that interweave and give force to those elements that are clearly and calmly observable. Indeed, Rivers's famous analytic divisions between magic, religion, and medicine are wholly rooted in observable practice and their relations to the social whole, smoothing out any fundamental differences in the process.[29] It is a sensible scheme, but in the hands of someone like Malinowski the polarity of the framework is often reversed, emphasizing not that everything has its place but rather that even a practice as rooted in empirical science as medicine is shot through with elusive, invisible, and (from the ethnographic viewpoint) nonsensical forces.

While neither Rivers nor Malinowski nor Murray served as a direct resource for Benjamin Christensen in his preparation for *Häxan*, the confluence of forces within the human sciences that gave their ethnographic (and, in the case of Rivers, clinical and experimental) research life was a discursive field that made an otherwise impossible film *possible.* Christensen's film mocks Murray in the same way that Malinowski openly mocks Rivers.

Häxan's debt to trends in the human sciences and the arts becomes even more apparent in the second half of the film, as Christensen is much more directly invested in drawing together otherwise disparate realms of science, magic, and religion. Overtly, he is seeking to use modern, empirical methods to provide an explanation of the persecution of "witches," but as should be abundantly clear by now, the correspondences between the objects of study and the forces in play in the film go much, much further than that.

By 1922, however, it was clear to audiences that the force of a threatening, invisible Other could take many forms. If anything, it was the very technologies invented to enhance life that further enabled this potentially threatening force. *Häxan* makes this point somewhat obliquely, but the continual visual reference to evil and mass death certainly indexes the film as a product of the immediate postwar period. The realized visualization of timeless, dark, invisible forces unleashed on humanity is even more apparent in Murnau's *Nosferatu*, visually rhyming the ancient evil force of the vampire with very modern images of hysteria and sudden, catastrophic death. Anton Kaes claims that *Nosferatu* is a "film about mass dying [that] answers Freud's exhortation to confront death openly as it lets us vicariously and safely experience what has been deemed unimaginable: one's own death."[30] This is certainly true, although we would amend this claim somewhat to emphasize that the objective of both *Nosferatu* and *Häxan* is to visually realize a force that can potentially annihilate life rather than express a metaphysics of what happens when you die. Certainly death itself remains unimaginable in concrete terms in both films, but crucially the force of this ultimate invisible Other does not. We are shown vampires and witches and (very importantly) doctors, scientists, and authorities (secular and religious) operating within a vector which folds each back on to the other across time and space, bringing this *bad object* into partial view. Thus, able to finally, fleetingly, glimpse the force that gives rise to the witch or the vampire (or the hysteric or even the modern soldier), it is abundantly clear that *the bad object stays bad* and is in no way safe. In its attempt to realize the bad object of the witch, *Häxan* speeds toward a conclusion decisively marked by swirling elements within the vector of time and space that was 1922.

FIVE

Sex, Touch, and Materiality

The figurative expressions used in the Bible to convey truths are not lies.

—ST. THOMAS AQUINAS, *Summa theologiae* (1265–74)

The Devil uses them so because he knows that women love carnal pleasures, and he means to bind them to his allegiance by such agreeable provocations. Moreover, there is nothing which makes a woman more subject and loyal to a man than that he should abuse her body.

—HENRI BOGUET, *Discours des sorciers* (1610)

Possession is the sadist's particular form of madness just as the pact is the masochist's.

—GILLES DELEUZE, "Coldness and Cruelty" (1967)

Christensen, having dazzled the viewer with a visualization of the witch stereotype of the sixteenth century in the film's previous section, moves in his fifth chapter to a seemingly more familiar cinematic approach. Focusing on the interrogation under torture inflicted upon the Young Maiden—*herself* arrested as a witch after Maria's vengeful accusation during her trial (because all witches know one another)—the film now highlights the intrigue, manipulation, and underhandedness of her inquisitors. Interspersed within this personalized narrative are Brother John's struggles with his sexual desire for the Young Maiden. Thus, while this section of the film expands Christensen's thesis regarding witchcraft and its relation to psychological states, illnesses, and diagnoses, the mode of presentation shifts to accommodate a more explicitly melodramatic style. Although skillfully rendered, it is fair to say that this chapter is not as innovative as the others in regard to its cinematic technique. The importance of this section of the film should not be underestimated, however, as its somewhat more conventional, theatrical style allows Christensen to carry forward his thesis regarding what the criteria for felicitous evidence had become in the context of the witch craze.

At this stage in the film, the reality of the witch becomes more explicitly multiple and simultaneously subtle. Moving away from a furious presentation of the witch stereotype, the interpersonal melodrama raises questions regarding the power of touch and the status of bodies that are explicitly understood to be *virtual*. Although essential to Christensen's realist aim in presenting examples of what inquisitors and officials at the time understood to be admissible evidence, the complexity of the virtual body folds back on to the film itself at this juncture. In precise terms, *all* the bodies in Christensen's *Häxan* are *virtual* bodies. Thus, building from Aquinas's linked assertions that the parables of the Bible nevertheless teach us something *true* and virtual bodies of angels and devils are *real*, *Häxan* self-consciously extends this claim to incorporate its own relation to the empirical. In keeping with the naturalism of *Häxan*'s style, Christensen begins by again populating his shots with clichés, but this time with those of romantic melodrama rather than supernatural proto-horror. And, as before, by the conclusion of this chapter the director will have emptied the screen of these clichés, locking in the relation between sex and materiality that has been established in the previous sections and, in the process, opening the door to a critical engagement regarding the very nature of a "real" body as it is contained in the image. Crucially, in emphasizing the centrality of sex as evidence for inquisitors in conducting witch trials, Christensen moves from his focus on a relationship formed freely through a pact and solemnized through sexual acts to violently totalizing acts of possession.

Contagion

Chapter 5 of *Häxan* opens with a series of title cards asserting that Maria's confession has set off a chain of events that were inevitable during the witch trials of the sixteenth century. As the inquisitors are successful in gathering evidence against the accused, the network of members in the cabal of witches that were to have existed becomes apparent. In all likelihood, Christensen was aware of the controversial character of the roving inquisitional teams that went from town to town in Germany and elsewhere in Europe during this time; considering the influence of the *Malleus Maleficarum*, he appears to be referring to the violent social disruptions directly caused by its author Henry Institoris (Heinrich Krämer) during his time as an inquisitor in the 1480s.[1] What Christensen does not show the audience is the stout resistance to this kind of persecution and upheaval demonstrated by civil authorities and some members of the clergy. Nor does he explain that often (as Institoris's case demonstrates) inquisitors were criticized for precisely the blinding zeal and the tactics depicted in the interrogation of the Young Maiden shown in this chapter. Christensen offers a streamlined version of "the witch" in presenting his cinematic thesis; here, the generally effective strategy is occasionally pushed beyond popularization and takes the film into moderately deceptive territory in relation to its subject.

Thus, Christensen's characterization of the scene of the inquisition is only partially "correct." The assertion that demonological beliefs were widely accepted within both scholarly and popular discourses is true. From the vantage point of the early twentieth century, this fact appeared to ratify claims that the period was one where naïve superstition circulated as a kind of misguided, yet common, sense. Yet *Häxan* gives no real indication at this point as to how, to paraphrase Alasdair MacIntyre, doubt, skepticism, and opposition from within this style of reasoning is critical to our historical understanding of the period.[2] As Stuart Clark has shown, by 1600 there were strong efforts to undermine the conceptual basis of witchcraft beliefs, refute the idea that witchcraft could be understood legally to be a crime, and criticize the severe social disruptions and potential injustices likely to occur within the viral movement of the witch hunts. Johann Weyer's *De praestigiis daemonium*[3] was a landmark volume in the growing corpus of skeptical writings for its attempt to extend the concept of delusion to *all* elements of witch confessions. In the coming decades others would follow, from Tanner and Meyfart to Thumm, to name but a few.[4] It is important to note that none of these works fundamentally challenged the overarching theological logic of demonology or the style of reasoning it upheld. Rather, based on combined appeals to theological conservatism (through a different take on Aristotelian method and harking back to the *Canon Episcopi*) and a proto-humanist concern over the injustices resulting from potentially erroneous confessions, these theorists suggested that witch trials should be resisted because they constituted errors according to the very demonological logic that guided their resistance to a transcendental evil. Christensen therefore provides the viewer with an indication of how violent and disruptive witch trials were and clearly marks these events as the material signs of the discourse that authorized them. What the director does not sufficiently demonstrate in *Häxan* is the diversity of opinion that existed within this discourse and the concrete ways in which elites and everyday people alike resisted the trials and worked to mitigate their often-catastrophic effects.

Bearing this criticism in mind, the chain of events that *Häxan* shows in this chapter have a historical basis and represent precisely the outcomes that troubled many, even in the sixteenth century. Staying with his case study, Christensen returns the viewer to the Printer's household, opening with shots of Jesper's wife, the mother, and the Young Maiden still tending to the stricken man. The magistrates who earlier arrested Maria are shown quietly sneaking in to the house through the kitchen. The mother is in tears; the implication is that they are now grieving the death of the Printer, although the film does not spell this out. The servant Sissal enters the kitchen and is immediately collared by the magistrates. The mother, hearing the commotion, rushes in and is also grabbed by the arresting authorities. Christensen's editing suggests a repetition of what happened with Maria, using largely the same camera setups as before.

Anna, the Printer's wife, distraught and grieving, reacts with horror at seeing her mother and trusted servant shackled in the small wooden wagon used to cart Maria off earlier. Christensen emphasizes the horror of the scene by cutting to the faces of the mother and Sissal, heads of wildly flowing hair, their faces contorted by screams. Aghast, she struggles briefly with the men and then rushes back into the bedroom, dropping to her knees with a fervent petition to God. The men having hurried off with their suspects, this prayer is intercut with shots of her sister Anna, having been knocked down in the melee, struggling back to consciousness and attempting to rise. Anna's face is streaked with the grime of the floor. She stands and then stumbles toward the open door to the street.

Noting that to oppose the accusation of a witch was generally taken as tantamount to being a witch oneself, the title card matter-of-factly states that Anna's destiny is now "sealed." Moving to the interior of a torture chamber, Anna is shown suffering the same trials to which she had condemned Maria earlier. Strung up by the wrists, the young woman hangs in the middle of the

"Witchlike" in *Häxan*, film still (Svensk Filmindustri, 1922).

chamber as her torturers go about their work. The sense of routine is emphasized, as several of assistants are shown occupied with a game to pass the time. A title card reminds the audience that there are only two people remaining in Jesper the Printer's "haunted house": a crying infant appears, sobbing, followed by a glimpse of the child being attended to by a young, frightened girl.

Masochism and Voice

> The film [*Häxan*] is not just scientific and artistic, it is an ethical event.
>
> > *Film-Kurier* magazine (1924)

Informing us that "during the witchcraft era it was dangerous to be old and ugly, but it was not safe to be young and pretty either," *Häxan* fades in with a shot of cherry blossoms in full bloom. Brother John is staring out to the courtyard, framed by the edging of the window and the hazy limit of the blossoms. He sighs, closing the wooden shutter as he ducks back inside. This brief sequence shifts the tone of the chapter considerably, to a well-established visual cliché of unrequited love. The expected exposition of precisely who or what is on the youthful friar's mind is not long in coming, as the young man approaches his older colleague Johannes, explaining that his "thoughts are sinful" and asks for help.

Taken aback, Johannes accompanies his troubled colleague to his cell, regaining his stern composure. In keeping with *Häxan*'s internal logic, the more experienced priest does not offer to talk things out with the troubled younger man. Instead Johannes simply asks the younger man to "bare his body," which he does after a brief hesitation. Taking a small whip in hand (which is quite handily hanging on the wall near the door), the priest promises to whip the "sinful body" and "poor soul" of the wavering young man, effecting a kind of "faith healing" in the process. Brother John is now kneeling, stretched out and steadying himself with a small stool. The upright older man begins to deliver on his promise and whips the prostrate Brother John.

Using a similar superimposed special effect to the one deployed earlier with the Wild Ride, Christensen presents a close-up of the younger man's face, grimacing and crying out, with the medium shot of the two men while the act of whipping taking place. Thrusting his body violently forward with every crack of the whip, Brother John's facial gestures suggest a combination of ecstasy and pain. The explicit homoeroticism of this scene is impossible to ignore, amplifying not only the earlier gestures in *Häxan* that hint at the role of repressed desire but also inferring a reinforced sense of the depravity of the inquisitors themselves. The barely disguised insinuations of this series is pushed further when, the whipping quickly at an end, the young man fervently demands, "Oh Brother, why did you stop . . . ?" Johannes regards his tearful, spent younger colleague and tenderly bends down to comfort him.

What has the viewer just witnessed here? This is a more difficult question to answer than it seems. The simplest answer is that Christensen has shown a technique critical to the cultivation of piety and ethical practice for sixteenth-century Christians. Quite unlike modernist, secular reasoning on the status of pain and suffering, for Brother John the experience of pain through techniques we would now associate with torture would have been essential to the cultivation of a pious self.[5] Bearing the mark of an Aristotelian logic whereby experience at the outer limit of the self worked to shape the character of the soul within, the disruption of Brother John's lustful desires would have to be addressed through the production of somatic experiences giving him the tools to resist his own thoughts. Thinking of himself in these terms, it is not surprising that the young priest actively seeks out his older colleague to inflict injury upon him. As Judith Perkins argues regarding early Christian martyrs, openness to pain lay at the heart of what structured individual agency for these believers *as Christians*.[6] Brother John would have certainly sought to emulate historical figures revered as saints by the Church. In light of this fact, Brother John can be figured as an active agent seeking to cultivate the pious disposition expected of an instrument of God, a reading that also (unintentionally, we argue) allows the audience a peek at the ethical basis for torture in general, including the torture of suspected witches.

Christensen does not actively deny this particular reading of this scene. This is hardly, however, the interpretation toward which he appears to be cajoling his audience. Instead, the director emphasizes the obvious Freudian echoes the passage conjures, particularly in regard to Freud's association of ecstatic religious ritual with neurosis and practices of pious bodily mortification with sadomasochism.[7] But as Niklaus Largier has shown in great detail, despite the inscription of sexuality onto scenes like these, "medieval and early modern sources specifically *do not* establish any relation between the effects of flagellation on the body and soul and what is now termed 'sexuality.'"[8] Nevertheless, it is difficult for the viewer to see anything but a sexual passion on Brother John's superimposed face as he endures the whipping from his elder. The reciprocity eroticism demands is strongly evident on the older priest's face as well, indicating an exchange, albeit one where the real passions driving the act are displaced within the violence of the affair. The image is among the most graphic in *Häxan* and continues to qualify for this appellation even by today's standards. Fully invested in the overloaded realism of naturalist cinema, *Häxan* almost achieves the direct truth effect *for Freud's empirical reality* that Bill Nichols controversially ascribed much later to ethnographic film and pornography.[9] It does this via the affect generated through the passion of the character's gestures and the visibility of Brother John's open, suggestively vaginal back wounds. Lacking an intellectual commiseration with the meaning of pain for these priests, Christensen instead engenders an affective sympathy toward them, seeming to anticipate that disapproving censors would not allow him to go any further in giving Freud's religious, repressed neurotics a cinematic, virtual body. As it turned out for the censors, he had already gone too far with scenes such as this one.[10]

There are fissures in Freud's many accounts of sadomasochism, however, and such fissures are evident within the image of sadomasochism in *Häxan* as well. Although one finds varied accounts in Freud's writings, the popular gloss of his theory asserts that sadism and masochism are rightly coupled into a single term because masochism is a derivation of sadism through a process of reversal. Freud's claim here is based on duality he proposed between the sex and ego instincts.[11] This chain of psychoanalytic associations appears to neatly parallel with demonological logic that presumed the power of witchcraft as being a reversal of the logic of the sacraments. As we have already noted, this characterization is not precisely accurate, and while Christensen was clearly familiar with the writings of Freud, he was also likely familiar with Richard von Krafft-Ebing's popular *Psychopathia Sexualis*, which takes masochism and sadism as distinct categories of neuroses (a view even shared by Krafft-Ebing's critics, such as the physician and writer on human sexuality, Havelock Ellis).[12] Again with the sadomasochist, a simple duality does not quite hold up, as this scene from *Häxan* demonstrates.

The most obvious issue is that *there is no sadism here*. A characteristic element of the libertine in Sade is his elaboration of the demonstrative power of language, his victims figured as unwilling listeners and participants in acts of cruel passion intended to reflect the superior forms of violence to which the demonstration affirms.[13] Brother John is neither an unwilling participant, nor seeking to "learn" anything from his older colleague in submitting to the whip. Rather, the younger priest is faced with a desire that cannot be brought into expression in any other way. While the older man overtly initiates the flogging, Brother John would have certainly known what form of penance would result from his feeble, stumbling confession ("I am having sinful thoughts"). The senior friar is only fulfilling an implied contract of sorts here, providing a vehicle for Brother John to "speak the language of the torturer he is to himself."[14] Confronted by a sexual desire that is nevertheless undoubtedly his, it is *Brother John* who fully executes the

Brother John's flagellation in *Häxan*, film still (Svensk Filmindustri, 1922).

function of the torturer here, demanding to express himself in a way that not only results in his own obligatory flagellation but also the torture and eventual death of his object of desire, the Young Maiden. Thus, the young friar in *Häxan*, far from being the naïve victim Christensen figures him as, exists as an agent more consistent with characters in Leopold von Sacher-Masoch's pornological fiction than Freud's descriptive typologies. The character of Brother John ultimately confirms the *indivisibility* of the masochist that was intolerable to Freud's science.

Christensen is also playing a somewhat dangerous double game in this scene in its expression of a barely concealed homoerotic desire between the two priests. Scandinavian film in general at this time exhibited a complex relation in its portrayal of same-sex desire that was ahead of its time. Depictions of this desire and the intricate affairs that they often generate in a wider milieu of repression and disapproval are dealt with openly and in a manner that, while still often referencing clichés, nevertheless displayed a compassionate realism when compared to other contemporary examples. Christensen would soon play such a role himself in Dreyer's *Mikaël* (1924). In bringing a convincing depth to the character of painter Claude Zoret, in love with the titular male model, Christensen's sympathetic acting is a real accomplishment that resists the lisping mawkishness inflicted on nearly every gay male character in cinema right to the end of the twentieth century.

The scene of Brother John's submission to the whip is not markedly different in this respect, but it does conjure a set of associations unlike those arising from Christensen's role in *Mikaël*, notwithstanding their formal similarities in the use of compositional techniques associated with tableaux and faces discussed earlier.[15] First, the explicitly somatic focus strongly indexes the scene as solely one of repressed sexual desire; we do not know the characters in the scene well enough to intuit deeper emotions they may have toward one another. In light of our general argument here, locking in the primacy of sex as evidence for the real is internally consistent. However, it is also somewhat dangerous, as the implied erotic dimension of this scene is read in the context of Christensen's overt message that the priests are hypocrites and zealots in their pursuit of a strict, moralizing denial of sexuality. As the image surely must be read in the context of its time, it is very difficult to avoid the conclusion that the homoeroticism at this point in the film is being figured as a manifestation of a repressive illness and to some degree serves the purpose of amplifying the Freudian sense that the priests are neurotics—ideas and a descriptive language of which Christensen was keenly aware. Reacting to these signs with either "pity" or "disgust" for the priests is left entirely to the viewer.

The homoerotic coding of the scene can also conjure more obscure associations. If we extend the reading of this repressed same-sex desire on the part of priests, the logic of the scene leads us back to some of the most unpleasant associations that exist in the demonological literature. In particular, the collapsing of the witch, the Jew, and the homosexual into the singular category of the sodomite is brought to mind. This chain of associations was not universal within the demonological writings, but it was not uncommon either. The most explicit articulation of this synthesized category is found in Manuel do Valle De Moura's *De incantantionibus seu ensalmis*, published in Lisbon in 1620 and beautifully analyzed more recently by Armando Maggi.[16] Reflecting the particularly virulent anti-Semitism of the Catholic

Church in Portugal, De Moura designates the term "sodomite" to include any individual person who is understood to have given himself or herself over to Satan. This would include the full list above and was flexible enough to potentially include others, such as the supposed cannibals of the New World. Significantly, De Moura's focus was on the assumed perversion of Catholic idioms evident in the language and practices of so-called sodomites. In other words, the ultimate crime of the sodomite was the perverted *inflexibility* of what De Moura designated as their *ensalmus*, a concept used to denote any invocative act.[17]

Haunted by the specter of such a ferocious concept, Christensen's insinuated linking of the priests with the sodomite collapses the numerous discrete contexts latent in the latter's potential meaning. The lashing of Brother John is itself figured as an *ensalmus*: the insinuation of homoeroticism would likely evoke sodomy directly in the mind of the viewer. Buried deep within this association is a discourse that would link the sodomite to the hypocrisy and heresy of giving oneself over to a false law, denying the idioms of "freedom" and "morality" reserved for the proper language of Christianity. Speaking in the name of science, Christensen joins the hypocritical priests with the witch under the sign of *superstition*. The relay that acts to abut these subjects in this particular scene, once held as absolute opposites, is the invocation of sodomy in the course of executing what for the director would have been an infelicitous, false expression of the truth of the world and a "proper" ethics in the face of "objective," scientific reality.

The Possessing Touch of the Other

Faith is still a matter of passion, of affect and nothing else.

Gilles Deleuze, *Cinema 1* (1986)

Perhaps we have assumed something here, as careful viewers will note the scene of Brother John's lashing implies the presence of the witch but she does not actually appear. Thus, the chain of associations that we have suggested is not yet complete. As the next scene shows, Christensen does not wait long to complete this chain. Moving forward from Brother John's quarters, *Häxan* cuts to Father Henrik poring over a manuscript. Johannes enters, informing the senior inquisitor that Brother John is spellbound and that a witch has been tempting him in his cell. Father Henrik is, of course, shocked by this news. Moving to a visualized sequence showing the Young Maiden's tempting "visitation," Brother John is shown in close-up, eyes open wide in fear. Cut to a close-up of the Young Maiden, her eyelids heavy, her lips barely open in a look of erotic expectation. Tears run down her cheeks, but her facial gesture is explicitly one of desire. Appearing adjacent to the previous scene of highly charged homoeroticism, the Young Maiden's "visitation" of Brother John in his cell functions as a suggestive visual rhyme with the inexperienced friar's earlier agony/ecstasy. The inquisitor's cell is clearly a space of suppressed, dangerous eroticism in *Häxan*.

Still in the facial close-up, the Young Maiden coquettishly looks down. She and Brother John are now shown in a medium shot, the young man backing away from the forbidden, dangerous woman. He suddenly cowers down at the table where this confrontation is taking place. The Maiden touches him, but she slowly fades from the shot, a special effect implying that this is all in Brother John's mind. Christensen returns the viewer to the prurient older men discussing the young man's bewitchment. "And she has grabbed him by the wrist," Father Henrik is told, this information offered as apparent proof of the Young Maiden's lustful, *malefic* power over their inexperienced colleague. They go to Brother John, shown gingerly lifting his cassock over his bloody, ravaged back. He is worried that Johannes has given the Maiden away, a concern that draws a sharp rebuke from Father Henrik, entering with a formal accusation to be signed. Threatened with the charge of being in league with a witch, Brother John reluctantly swears out the accusation, the intense pressure he feels coming to life through a series of quick reaction shots showing the angry intensity of his superiors. The Young Maiden's fate is, indeed, sealed.

Viewed in an era after the appearance of ahistorical witch polemics by neo-pagans such as Mary Daly[18] and Starhawk,[19] it would be easy for today's viewer to interpret the meaning of this scene as a display of the blunt misogyny of the Church that would have ultimately caused the witch craze. As convenient as such a reading is within the identity politics of the present, it is undeniably *false* in light of what *Häxan* shows in the scene and in the context of the historical record. The indignities of the *Malleus Maleficarum* in regard to women are well known and useful to such a polemical reading.

Echoing Jim Sharpe, we claim that it requires more than just this sense of indignation to make such an interpretation ring true.[20] It is hardly accurate to assert that the *Malleus* was universally used as an authoritative source during the time of the witch trials and, as we have shown, the famous manuscript is not even the primary source for much of what *Häxan* depicts.[21] Furthermore, the scene shows what the *Malleus* and many other demonological texts *actually* claimed was the causal source of witchcraft. The cause was not women; rather, the source was *carnal lust*.[22] Within the logic of the witch, women were "by their nature" much more predisposed to giving in to "inordinate affections and passions," hence their related susceptibility to the power of the Devil. As odious as such logic is, it must be said that this is not the same thing as hating women. Christensen, therefore, does not show inquisitors who merely despise women. If anything, their own carnal desire for these women overwhelms the priests to the point that they pathologically fear them, displacing their own sensed alterity to themselves within this generalized feminine image. While the stereotype of the lustful, rebellious, rancorous woman was wildly inaccurate in its clumsy universalization and certainly harmful in an environment of deadly suspicion, it is a fact that demonologists showed little interest in exploring it as such, as a primary agent of witchcraft. Therefore, the focus in these scenes in the film is squarely on *desire* rather than *gender*, which is both conceptually consistent with the Freudian inspiration within Christensen's thesis and empirically sound in relation to the historical record itself. Much like Dreyer's *Day of Wrath* (*Vredens dag*, 1943), where the desires of a young woman unhappily married to a much older pastor spark suspicion and jealously, sexuality itself is the relay between the natural and supernatural—a relay that more completely implicates women but does not exclusively focus on them.[23] Paraphrasing Katharine Rogers, it is obvious that *Häxan* does not display wicked *women*; rather, it shows *witches*.[24]

There is something else in this scene. Strangely, the report given to Father Henrik does not state that Brother John is having "sinful thoughts" (although this is what he originally confessed), but instead claims "the witch visits him in his cell." Taken literally, the cinematic grammar of the sequence, seems at first to contradict the older friar's claim that the witch is anywhere near Brother John's cell, as Christensen deploys the special effect of having the Young Maiden fade from the scene like a menace from a dream. This effect was a well-established sign by 1922 that the spectator was viewing a hallu-cinated image. The audience is hardly confused by what they are seeing; interestingly, Brother John has come to the same conclusion and does not directly claim that the Maiden "herself" is anywhere near his cell. Why, then, does Johannes appear to either misstate or exaggerate the character of his younger colleague's torment?

Of course, this question presumes an agreement between the character, the filmmaker, and the viewer as to what the "actual" character of Brother John's torment is. In short, each subject position must hold that the image of a dream or hallucination is *not real* for the question to be a valid one. By now, however, it must be clear that there is a pronounced ontological arrhythmia that exists between characters, viewers, and the director in *Häxan*. In her recent study of Christian materiality in late medieval Europe Caroline Walker Bynum provides an example of what Christensen is at-tempting to synchronize in the film. She demonstrates that a theory of the image that would expel it from the domain of the materially real is inade-quate to the task of grasping this materiality. Rather, we must understand how the image, even an image known to originate in the mind, was not only real but *alive*.[25] In making this claim, it should be abundantly clear that we are not arguing that these Christians could not distinguish between the con-tent of their dreams and other forms of reality and life. To the contrary, the deep fissures that existed within the discourse of the witch demonstrate that there was a great deal of sophistication in the various designations of not only what was real, but also *how* something was real. Christian materiality in late medieval Europe, while not precisely *our* materiality, was neverthe-less quite concerned with determining difference and ascribing designations to discrete things.

It is obvious that the inquisitors are well aware that Brother John is re-ferring to an image that is tormenting him. This image bears a direct rela-tion to the body and self of the Young Maiden in that it originates in her and is projected outward, presumably through the supernatural relay of the power she supposedly possesses as a result of her pact with the Devil. The girl's visitations to Brother John serve as concrete proof of Satan's activity, as the Maiden could not do this alone. Crucially, the Devil could not by him-self reach a pious instrument of God in such a definitive way. Without the witch, Satan can only annoy, tempt, and deceive. The witch extends the scope of Satan's power on earth by lending her fleshy body to his virtual one, a promise literalized through their obscene sexual intercourse.[26] The Devil and the witch depend on each other here, as they together allow each other to unnaturally extend their respective material forms. In this specific instance, this is twice that she has unlawfully touched Brother John.

This is also a very *Protestant* way of understanding an image, particularly in reference to the relation between an image in thought and things in the world. Although Christensen is clearly depicting Catholic characters, the emphasis he places on portraying a concrete link between what Luther termed "an image of the heart"[27] and the reality of the divine or supernatural beings and forces that correspond to them, bears a close resemblance to early Reformation debates over the empirical relationship between an image and the inner message of Scripture itself. Luther is referring directly to the image of Christ in pictures inscribed in the heart, seeking to find a middle ground between iconoclasts such as Andreas von Karlstadt and Catholic "idolaters" in formulating a theory of the image. It stands to reason, however, that the danger of malevolent images of the heart would likewise correspond to their diabolic counterparts, making the image of the Young Maiden in Brother John's heart all the more dangerous and real. Christensen's scene therefore reflects (most likely unconsciously) the fierce debates that took place within early Protestantism regarding the character of what a supernaturally charged image actually is.

The Young Maiden's erotic presence before Brother John is not, under these conditions, simply a fantasy originating within the mind of the young priest, but a torment invading his very self—*a projection from the outside*. In this form, the Maiden is not a figment of Brother John's imagination, but an image that can *touch* him. The form of this logic is consistent with the rationality Bynum convincingly outlines regarding the ontological status of icons, relics, and other supernaturally charged images. Such special images simultaneously *are* and *are not* what they visualize within this style of reasoning, without contradiction.[28]

Easily unnoticed in the erotic swirl of the scene is the fact that what is happening has shifted somewhat. *Häxan* is no longer displaying to the viewer the terrible events of the Sabbat or the tragic outcomes of *maleficium* enacted through witchcraft. It shows us the ability of a "fallen" woman to penetrate the soul of a pious priest. She has truly gotten under his skin, a

The possessed in *Häxan*, film still (Svensk Filmindustri, 1922).

fact brought home to the viewer through the young friar's masochistic demand that he be flayed open with the whip. His colleague opens wounds but
does not drive her out. "Now my soul will surely be damned," Brother
John laments. But is this witchcraft? The viewer is led to unwittingly carry
over this interpretation, but the inquisitors do not actually designate the act
of the Young Maiden's visitation as witchcraft per se. Her spectral presence
does provide them with material evidence that she is a witch, but this is due
to the originary source of her power and not characteristic of the act itself.
In short, *Häxan* has shifted its focus from an account of witchcraft to a
phenomenon more precisely defined as *possession*.

   This shift is significant. In the increasingly mobile exchanges between
bodies virtual and substantial, the power hitherto ascribed to the witch is
starting to become detached from her. The danger evident in this fluidity is
no longer limited to that of the misfortunes resulting from *maleficium* but
now also includes losing oneself to another. Witches are lost to the Devil,
but the logic of the pact demands that they do this freely, acting in their own

name. *Häxan* is in different territory now. The Young Maiden has not "bewitched" Brother John, despite what the senior friar reported to Father Henrik; rather, she has *touched* him. Taken in light of a theory of images dominated by holy iconography that demanded to be touched, it is not a leap for Brother John's colleagues to take his visions as *visitations* and the Maiden's sexual aggressiveness in this virtual form to be concrete evidence of her status as a witch. In a certain sense, the "realness" of the Young Maiden for Brother John is not radically dissimilar from her status in relation to the viewer, as the object of a film viewer's gaze is in the end received as a "flesh and blood reality" as well.[29]

At its most concrete, every *body* in *Häxan* has the relation of the Young Maiden's-body-to-Brother-John or Satan's-body-to-Maria. They are bodies simultaneously flesh and image. The formal elements of Christensen's rendering of the scene between the voracious virtual body of the Maiden and the cloistered boy (Brother John is, in the end, nothing more than a "boy" in this sequence) rupture whatever hope the audience may have had that *Häxan* would somehow either entertain or inform them in any unproblematic way. Yes, the Young Maiden's visit to Brother John is marked as a hallucination. Yet the affect the scene produces, mainly through the erotic certainty of the Maiden's tightly framed face, is that she is *there*. She is staring not at Brother John but at *us*, reversing the cutting, surgical gaze Walter Benjamin attributed to the camera.[30] How, under the demand of hard objectivity, can any film show "reality" within these parameters? There is never more evidence of Christensen being himself "caught" by the witch than the Young Maiden's close-up in this scene—it qualifies as *truthful* under precisely the same ontological assumptions on display as a *problem* in the film. In rendering the Maiden so erotically desirable and so *close*, Christensen has his audience right where he wants them. Torn between the *belief* in a film image's "illusory" status and the *sense* that what one is seeing is as real as anything else, the audience can only retrospectively and discursively put a distance between itself and the image before it. An avowed pact with the Devil being largely impossible by 1922, Christensen nevertheless demonstrates the power of the witch that has ensnared him through an instrument more obviously available to him—possession.

The shift from an emphasis on those who have made pacts that give them voice to individuals who are potentially victims of invading, unwelcome

The possessor in *Häxan*, film still (Svensk Filmindustri, 1922).

others returns us to our earlier discussion of how Christensen has innovated and subverted (most likely unintentionally) Freud's notion of the "sadomasochist." Christensen has been leading the viewer to conclude that the authorities that conducted witch trials were violent oppressors in relation to the accused women. More directly, their sadism in pursuing witches is essential to the audience's sympathy for the "objective" diagnosis of what "really" happened later. Yet, in their own insecure efforts to demonstrate, to educate, and to persuade, these inquisitors do not meet the definition of the sadist at all. The sadist would not attempt to justify or demonstrate the felicity of his actions; he is *certain* that any reasoning is a form of violence and comes out *unequivocally* on the side of that violence.[31] The entire discourse of the witch undermines any such certainty. In fact, if anything the witches and (particularly in the case of Brother John) even the inquisitors tend to more closely resemble the masochist; this is a serious problem for Christensen's thesis. He partially resolves the issue in this chapter of the film by attempting to forge a visual equivalence between the witch and the possessed. While blatantly violating established demonological categories, the narrative strategy works in that the possessed more logically occupy the status that *Häxan* requires.

There is an unintended consequence to twisting demonological categories as a narrative strategy in the film. In doing so, Christensen has shifted the power of the witch back to the Devil himself. Satan is, after all, the quintessential sadist. The move appears fluid within the logic of the film, yet it has no actual place in the discourse that Christensen sought as a resource. Like the devils demonstrating their power through their possessed victims, Christensen is now seizing the *idiom* of science in order to express a truth through his own style of reasoning. When we move to the depictions of obsession, hysteria, and illness in the final two chapters of the film, it is critical to bear in mind that the filmmaker has by this time ventured so close to making a case for witches and witchcraft that he can only depict her through images and language already possessed by her power. Christensen, at this point in film, must rely on the witch (her reality, her presence) to make his case.[32] The tragedy of this section brings a logical coherence to *Häxan* that its *Mnemosyne*-like style in earlier chapters had resisted. Whereas *Häxan* previously identified who or what the witch was, now the film identifies *with* her. The coherence Christensen creates at this point resonates more precisely with the world of the witch than with the filmmaker's claim at the outset, to dissect, analyze, and dismiss the witch as an in-credible phantasm. Now it would seem the witch is an instrument of knowledge and the expression of supernatural reality.

The Late Arrival of the Sadist

The Young Maiden's danger to the inquisitors is obvious from their point of view. Having provided some context for their vicious response to her, Christensen moves to showing the torment of the Maiden's interrogation. Introducing more elements of the witch stereotype, she is confronted with the awe of the Divine through His instruments in the Church. Unable to cry on cue (a sure sign of being a witch), and having endured the tortures depicted previously (the sequencing here implying that the Maiden was under arrest for witchcraft well before Brother John's forced accusation), Father Henrik declares that these "typical" strategies of eliciting a confession will not work on her. The remainder of this section is taken up with the unusual lengths to which the inquisitors will go in order to gain the confession upon which they vitally depend.

In short, the inquisitors plan to entrap the Young Maiden, offering her freedom in exchange for knowledge of the "beautiful art" of weather magic, something well established long before the advent of the witch.[33] In a variation of the "good cop/bad cop" routine deployed earlier in the interrogation of Maria, Johannes frees the Maiden from the stocks in which she had been imprisoned. The title cards imply that she has been in this excruciatingly painful device for days, and she is shown to be unconscious from the pain of the ordeal. Liberated from the stocks and brought around with a splash of water, the Maiden is dramatically offered her freedom by the friar. Father Henrik and his subordinates spy on the scene out of sight. The Young Maiden is no fool; she accuses the monk of "taunting" her and flatly refuses to show him how to "make thunder" with the bucket of water that was just used to bring her back to life. Brother John has also slipped into the dungeon and watches this scene, curious and anxious, from a different concealed position than the one used by the others. Father Henrik, given the signal from Johannes that they must intensify the pressure, leads the team of inquisitors, magistrates, and guards in a pantomime of picking up and leaving town, lending credence to the offer of freedom. The Maiden, struggling physically against the monk as he tries to force her over to the water, still refuses, claiming no knowledge of weather magic.

Intensifying things further, the malicious inquisitor remarks that Anna and Jesper's child is now alone in the world and will die unless the Young Maiden is able to gain her freedom. Christensen's familiar technique of sequencing matching shots manifests the terrible intensity of this exchange. The Maiden, bruised and brutalized, her hair cut short, and obviously terrified, is broken at last. Father Henrik and his entourage, now inside, produce the child for the Maiden to see (and roust the objecting Brother John, now discovered). The Young Maiden, unable to bear this any longer, tells a story relayed to her by an itinerant stonecutter regarding how dipping one's hands in the water can produce thunder. She does not directly admit witchcraft or claim to have ever used this technique herself. Before she can

The Young Maiden under torture in *Häxan*, film still (Svensk Filmindustri, 1922).

even demonstrate, however, Father Henrik triumphantly makes himself known, shouting that the Young Maiden will burn for being a "hardened witch." The gravity of what has happened sinks in and the Maiden whirls around to strangle Johannes, who has so explicitly lied to her. She is restrained by the guards and bundled off, her "confession" now complete.

The arc of the Young Maiden's interrogation here is different from what we witnessed earlier with the torture of Maria. The Maiden refuses to give voice to the scripted pact the inquisitors expect her to ratify, which distinguishes her from the old woman who, after a "minimum" of bodily trial (if we can crudely put it that way) almost eagerly takes up the formulaic libretto of the lethal opera in which she found herself cast. The cheap tricks and almost laughable standard of evidence displayed by the interrogation of the Maiden reflects the profound shift in *Häxan* described earlier. The sadistic inquisitor has now arrived, but Christensen has not yet firmly established the ground for the character. Briefly slipping into a rudimentary understanding of Freud and the witch craze alike, it is clear that Christensen will have

Unknown artist, *The Sorceress* (1626). Courtesy of British Museum, London.

to do better than this to forcefully bring *Häxan* to its conclusion. Thankfully, it does not take the director long to regain his cinematic footing within the shifting terrain of the film.

The film's chapter ends with the broad statement that "the witch madness, like a spiritual plague, ravages wherever these judges go." Interspersed between shots of the full inquisitorial party packing up and moving on (almost identical to the ruse for the Maiden's benefit), Christensen claims that over eight million women were murdered in the course of the witch craze. This is a wildly erroneous figure that was propagated by early historians of witchcraft such as Étienne Léon de Lamothe-Langon and Joseph Hansen, and the director uncritically reproduces the number here. Christensen could not have known that Lamothe-Langon's figure was based on a fabrication, but the ridiculous inaccuracy of the claim is still jarring.[34] Regaining some sense of where *Häxan* is going, and putting this fancifully inflated inaccuracy aside, the chapter draws to an elegiac close, with the inquisitors crowding through the gates of the town, moving on to their next site and their next victims.

The sight of the inquisitors marching through the town gate cannot help but recall the fresh memory of invading troops familiar to European audiences in 1922. Anton Kaes has argued that the Great War deprived cinema of its ability to move freely, an observation that holds in light of our earlier discussion of Christensen's largely immobile camera in *Häxan*.[35] This restriction of mobility has another sense in this scene, however, in that the space reserved for Satan and his accomplices is increasingly being colonized by *institutions*. As in Dreyer's *Day of Wrath*, the Church is identified with a coldly dispassionate law, victorious in its repressiveness and institutional contagion.[36] Now free of their cloistered spaces, the shock troops of God and the Church annex the wondrous space of the demonic, appropriating the satanic power of the witch for its own ends. It is *startling* to suddenly see the priests outside, moving about in their endless search for the witch. For Christensen's purposes in *Häxan*, the scene is a particularly effective one in giving fuller expression to the phenomenon of possession that overtakes the witch in the following section of the film. It is this dynamic of possession to which we must now turn.

SIX

Possession and Ecstasy

> Convulsive flesh is the body penetrated by the right of examination and subject to the obligation of exhaustive confession and the body bristles against this right and against this obligation . . . [an] involuntary revolt or little betrayals of secret connivance.
>
> —MICHEL FOUCAULT, *Lectures on the Abnormal,*
>   *Collège de France* (1974–75)

The film's sixth chapter begins with a statement that leaves little room for misunderstanding: "There are witch confessions that are totally insane." At first the penultimate chapter of Christensen's thesis seems ready to offer more of the same (dramatizations of accusations, inquisition, and punishment of witches), now with the introduction of "insanity" to deepen of the film's attention to possession and demonic influence. But here Christensen instead shifts into another mode to further his cinematic thesis. His purpose is not only to show that witches arrived at their confessions thanks to more than a little help of the inquisitors; he also wants to record for the viewer the mechanisms and creative instrumentation developed to generate such confessions. In addition, he wants to give proper attention to confessions that were not *procured* but rather *volunteered* and *enacted*—expressed without the aid of the inquisitors. These were confessions of another sort. They were *possessions*—some spontaneous and others voluntary or as pacts with the Devil—manifested not from juridical manipulation but out of individual turmoil. In the chapter, Christensen moves away from the historicization of his subject and instead concerns himself with its medicalization—a mode that is equally didactic and forensic. With *Häxan* this move occurs in three distinct ways: first, through testimonies that fit into the genre of transfiguration and self-confession; second, through a close analysis and showcase of instruments of torture;[1] and finally, through the highly charged expression of religious ecstasy and self-mutilation as part of visual tableaux from his major source material: late-nineteenth-century works in neurology found in the writings of Jean-Martin Charcot and his followers, which explicitly dealt with the relation between witchcraft and nervous disease.

Transformations

Although spectacular images of hysteria and nervous disease have been consistently gestured toward in the composition of shots up until this point, the specter of mental disorder has not been explicitly raised in the film. The insistence on "insanity" by Christensen is, in this case, of a particular type—namely, the disposition for causing harm to oneself or acting out in a way that is guaranteed to be considered blasphemous. The film focuses on one category: "insane confessions," which are grounded in transfiguration. Christensen suggests that there are many such accounts, specifically of witches claiming to transform into cats in order to sneak into churches to defecate on altars. In the opening sequences of Chapter 6, we find hooded women (presumably acting as cat-women) entering a church, as large hybrid human–animal demons (pig-like creatures) stand as sentinels at the door. No specific citation is offered here, but Christensen clearly has some reference in mind. The scene is gruesome but it should be noted that the costuming is not one of *Häxan*'s finest moments. While the outfits do bear some faint resemblance to medieval depictions of human metamorphosis, this likeness is generic, the closest model being the animal-headed, human-bodied creatures that appear in woodcuts from Lucius Apuleius's *Von ainem gulden Esel*.[2] The moments when Christensen chooses not to make direct reference to a single historical source in his staging is interesting only because it so sharply contrasts his attention to "accuracy" in other scenes. This aside, Christensen seems to be making an important distinction; while all witches's confessions are mad, none are quite as mad as the ones claiming metamorphosis.

In a state of possession, witches would engage in a range of aggressive acts, including performing rituals to destroy pregnancy or otherwise pollute the domestic sphere—activities that, even without denial, would lead to painful interrogation techniques to discover *how* one came to be possessed (there was always room to implicate others) and to what extent the *maleficence* had reached. In an early scene in the chapter, we find a demonstration of the belief that "tying knots" in a string over the marriage bed would cause impotence, or in Christensen's interpretation, destroy pregnancies, each knot representing a miscarriage or a moment of impotence. While these practices were very much a part of the repertoire of the folk magician of the time, Christensen is correct to note that such practices, generally not held to be an indication of full-blown witchcraft prior to the sixteenth century, could very well land its practitioners before an inquisition during the time of the witch craze. The director does not index this association as the definitive evolution of the stereotype of the witch, but he is safe in suggesting that this practice (along with many others) historically associated with popular magic is by this time sufficient evidence of witchcraft.[3] Interestingly, in *Häxan*'s visualization of the miscarriage curse, the actress who portrays the mother is also shown here to be one of the old women seeking to inflict harm on the absent couple that shares the marriage bed.

Instruments

Christensen is not satisfied with restricting his analysis to old women casting spells with the aid of "witch's hair" and "metal crosses." We return to the torture chamber where Maria was interrogated. Using precisely the same footage as before, we are reintroduced to the tools of torture seen only at a distance before. The scene is complete with a repetition of Erasmus the executioner slowly ambling over to the wall to collect a set of terrifying tongs. This time, however, the film lingers seductively in a close-up of the tool's handle, then slowly pans down the length of the tool in order to give the audience a proper look, indicating that it is no longer Maria but the tool that is the star of this scene. The camera rests on the sharp pincers, now opening and closing, emphasizing the potential brutality of the instrument. The audience sees a woman (possibly the Young Maiden), naked with her back to the camera, held in place with her arms behind her head. The lighting of the scene, which illuminates her armpit and a hint of her breast, emphasizes her exceptionally vulnerable position. The horrifying pincers come into frame, clamping down on the exposed flesh of the woman's arm; she crumples in pain, still held upright by the rough hands that bind her from the back. Lashed by her wrists with leather straps, a disembodied "masterly" hand crushes the woman's softly lit, naked flesh with the pincers. Christensen's composition of this scene unmistakably refers to literary sadism and genres of violent underground pornography.

The film's textual narration immediately following this sequence reinforces the doubled elements of the film's action: "You and I would also be driven to confess mysterious talents with the help of such tools. Isn't that so?" The tone, while referring to the fact that one would say anything to avoid the pain of torture, is vaguely suggestive in its reference to "mysterious

A tool of torture on display in *Häxan*, film still (Svensk Filmindustri, 1922).

A tool of torture in use in *Häxan*, film still (Svensk Filmindustri, 1922).

talents," playfully offered as a question. Coupled with an image that works to elicit desire more than sympathy or horror, Christensen's attitude here is ambivalent to say the least.

Christensen presents a catalog of the tools used to elicit confessions, each forged by one of his set designers to match original implements of the period. Christensen goes to great lengths to create "real" instruments—he wants his viewers to *dwell* on their purpose and precision. He takes careful inventory of these devices, and in doing so the audience is invited to think about the clever ways they can be used to elicit confessional speech. The scene implicates the viewer in the imagination of torture. The exercise also produces objective *distance*, as if the director is only turning the pages of an encyclopedia of scientific instruments. There is a cold exactitude and disinterest that stands in contrast to the panic and confusion of earlier scenes.

As the instruments are displayed, the camera turns dramatically away only seconds before the torturer puts them to use. The scene moves to a pair of feet bound to a table, panning up the body of the female captive to demonstrate how the accused was held in place as the torturer did his work. Strikingly, as the camera again pans up the body of the tortured, the model in this scene is revealed to be male, yet no explanation for this switch is offered. Following the large chain that binds the man's hands, the shot freezes briefly on a pulley. After a pause, the panning continues, this time moving vertically, following the chain upward to the mechanism that provides the necessary leverage to hoist whomever is anchored to the table. The torturer's arm glides through the frame, his hand menacingly gripping the handle of the contraption. Christensen again spares us the gore, moving instead to another scene, now of a foot-crusher, temporarily empty.

We have come full circle to the lecture style used in the opening chapter as a pointer enters the frame at various moments to direct the viewer's gaze. Christensen deflects the horror of these scenes by using stylized historical images, perhaps anticipating the disapprobation of audience and censors, or simply to remind viewers that these tools were born from an imagination that is not his own. The dark overtones of this sequence nevertheless remain intact as the scene moves to a woodcut "which speaks for itself" (credited to the "French Doctor" Régnard) titled "After the Interrogation." The woodcut depicts an unconscious, perhaps dead, victim surrounded by his inquisitors and the torturer, whose face unmistakably resembles Satan's. The victim in this image appears male, a fact that again elicits no comment.

Next the audience is informed that "painful interrogation" often began with a "light" torture. The scene moves immediately to a thumbscrew held up to the camera. As two feminine hands are clasped together, the woman's thumbnails are shown in the center of the frame. In a gesture of comical overkill, the out-of-frame "demonstrator" helpfully traces the cuticle and eponychium of the actress's thumbs for the viewer, emphasizing precisely where the pressure of the thumbscrew will be applied while simultaneously amplifying the sense of a clinical presentation. Christensen explicitly ruptures the pretense that we are witnessing something "in the past" by informing us that one of his actresses insisted on having the thumbscrew tried on her. We are told that she "actually" allowed the device to be screwed down, boring into her thumbs. At first smiling, the actress begins to shout and Christensen drolly informs the audience that he "will not reveal the terrible confessions [he] forced in only one minute of torture." The effect is one of a peculiar verisimilitude. We are led to believe that this was not an acted sequence but rather something done to simply "try out" the instruments. This brief puncturing of fictive barrier between "acting" and "doing" only highlights the constant ambiguity between these two realms that constitutes an essential character of *Häxan*'s formal cinematic strategy, aided by the film's oscillation between "fact" and "reenactment" in its presentation.

The Problem of Ecstasy

The demonstrative mode of *Häxan* abruptly shifts, and for the moment returns to reenactment. The audience is taken to the late medieval convents, marked (as we are told through intertitles) by "an almost hopeless despair." A nun looks directly into the camera, raising her hands in prayer. Mirroring the masochistic eroticism of the earlier scene in which Brother John is whipped, the nun reaches for a spiked belt she will wear as a form of "regrettable self-punishment." We see her go through the motions as she prepares. Apprehensive, tears flowing, she gathers the nerve to bear the pain she is about to inflict on herself. She cinches the belt up tightly, driving the spikes into her own abdomen. Christensen returns to her face, her expression a mix of pain, pleasure, and relief.

Christensen had originally planned to make two sequels to *Häxan*—*Saints* and *Spirits*—forming a trilogy that would survey religious thinking and the rise of modernity.[4] With *Saints*, he had hoped to explore the topic of religious hysteria—the violent visions which emerge at different points in history by pious church parishioners or clergy, often resulting in acts of self-harm, malevolent violence, or religious ecstasy. In the scenes of self-punishment in Chapter 6, we find Christensen setting the stage for his unrealized film. This, however, is in no way a departure from the thesis Christensen follows in *Häxan*, and in fact closely follows the cues found in the *Bibliothèque diabolique*—as cited earlier, a series of published studies organized by Charcot's student Bourneville and others that so greatly influenced Christensen—which explicitly dealt with the relation between religious ecstasy, magic, witchcraft, and "nervous disease."

Christensen informs his viewers that it only took one nun to be "seized" for the entire convent to be "overtaken with insanity—a mysterious, contagious insanity." The director offers a long look at a group of nuns in a church

Pierre Andre Brouillet, *A Clinical Lesson at the Salpêtrière* (1887). Depiction of Charcot and his students. Courtesy of Leonard de Selva / Corbis.

Illustration of Jean-Martin Charcot using light to hypnotize a patient. Courtesy of Stefano Bianchetti / Corbis.

dancing wildly to visually amplify the words on the title card. Christensen claims that the scenes that follow are derived directly from the surviving writings of nuns afflicted with this torment. Although he does not specifically cite any individual works through his title cards, we can be sure that these scenes come from *La Possession de Jeanne Fery* and other accounts in the *Bibliothèque diabolique*. Again, Christensen's choice of words here is significant, as he is setting the audience up for a scene of "how the Devil penetrated the convent" (*inträngande*), a phrasing that conjures the multiple, eroticized elements of the Devil's interactions with humans in the early modern period.

The focus turns to a nun gazing upward in wide-eyed supplication. Although her expression is obviously troubled, the visual echo of Charcot's (Régnard's) photography of Augustine in a state of *supplication amoureuse* is unmistakable.[5] Her terrified gaze is explained as the Devil towers over her, flicking his tongue lasciviously at the petrified woman. The intensity of the woman's expression hardens as Satan reaches into a wooden box that sits on

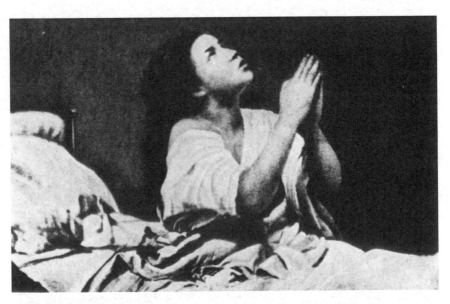

"Attitudes passionnelles: Amorous Supplication," *Iconographie*, vol. 2, from Désiré-Magloire Bourneville, Paul Régnard, Jean-Martin Charcot, and Édouard Delessert, *Iconographie photographique de la Salpêtrière: Service de M. Charcot*, 3 vols. (Paris: Progrès Médical, 1877–80).

Supplication of nun in *Häxan*, film still (Svensk Filmindustri, 1922).

the altar. His long, talon-like hand removes a knife from the box and offers it to the panicked nun, her movements exaggerated to suggest a near-paralyzed state, again a visual diagnostic derived from Charcot called "Tetanism."[6] Satan, gripping the knife firmly by the blade, forces it on the helpless women. Her body seems to "snap" as she reacts suddenly. It is unclear if she is begging for mercy or simply attempting to refuse Satan's demand. She shouts, "Get thee behind me, Satan!" but Satan, now shown in frame with the nun for the first time, merely fades away, as Christensen again deploys superimposition to convey mystery.

Overcome by Satan's power, the nun rises with the knife, now explicitly reminiscent of classical images of the hysterical woman. The nun throws away the offending knife in horror. Satan, who has faded from view, suddenly pops up from behind a small pulpit around which the action is taking place. Satan stands behind the nun, who is still reeling from her revulsion. She does not see the Evil One who takes advantage by violently smashing her over the head with a large club that he has produced out of thin air.

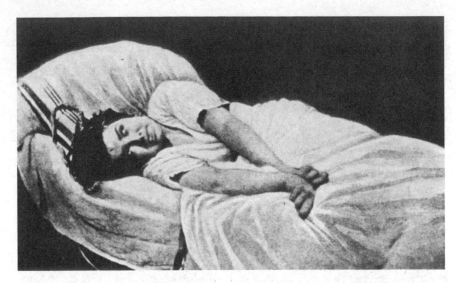

Augustine, "Tetanism," *Iconographie*, vol. 2, from Désiré-Magloire Bourneville, Paul Régnard, Jean-Martin Charcot, and Édouard Delessert, *Iconographie photographique de la Salpêtrière: Service de M. Charcot*, 3 vols. (Paris: Progrès Médical, 1877–80).

Nun with "Tetanistic" response to Satan in *Häxan*, film still (Svensk Filmindustri, 1922).

In an uncomfortable slapstick, the nun falls to the floor. The melodrama is heightened as the nun bounces back from the blow, her body erect in protest. Satan regards her with a warning, his tongue still flicking at the unfortunate sister, again raises his club as a renewed threat and she relents, grabs the knife, and stumbles to her feet. By the time she is upright, the Devil has dissolved from view, reappearing at the door at the far end of the room, beckoning the nun forward.

This violent action has taken place in a sacristy adjacent to the main part of the church. Satan has demanded that the nun, knife in hand, enter and approach the altar to carry out his demands. The film shows her progress through a series of overlapping dissolves, emphasizing in turn the intensity of her experience (in close-ups) and her diminutive status in relation to the forces in which she is caught (through long shots of the sister making her way to the altar in the cavernous church). Christensen then draws the viewer's attention to the open Bible on the altar; our view is identical to the nun's vision, as several reaction shots make clear. Satan suddenly appears among the objects on the altar, beckoning the sister forward. Making her way to an ornate wooden tabernacle, the knife still in hand, the nun extracts a consecrated host and places it on the altar. The host is shown in close-up, followed by a cutaway shot to the bas-relief sculpture of the Crucifixion over the altar. Christ appears as a ghostly apparition to the stricken nun, pained by what he is witnessing. Provocatively, as with Satan, Christ is played by Christensen.

The audience is momentarily denied a resolution to the scene, as the film leaps to the other nuns in the convent entering the chapel, at first unaware of what has been taking place there. As they enter, presumably to worship, Christensen shows the aftermath of the intense encounter between Satan and the stricken sister. Slumped against the altar, her eyes rolled upward and staring in blank incomprehension (again, explicitly referencing Charcot's diagnostic photographs of *attitudes passionnelles*), the group finds the possessed nun, arms serenely folded in her lap, bewitched and ecstatic. The knife, still grasped in her left hand with the blade downward, protrudes away from her lap, the sacramental body of Christ impaled on the sharp tip. In this carefully composed shot, the phallic overtones evoked by the position of the knife are unmistakable. Satan, through the necessary actions of the unfortunate nun, has raped the body of Christ.

Nun penetrates the body of Christ in *Häxan*, film still (Svensk Filmindustri, 1922).

The group of nuns rushes forward to examine the catatonic woman. The older nun realizes what is going on: "Sister Cecilia is conniving with the evil one." They rush away from the stricken woman in terror to the far side of the chapel. Sister Cecilia sees their flight and rises to her feet, knife in hand. The film cuts between a medium shot of Sister Cecilia struggling to regain her wits and medium close-ups of the faces of her fellow sisters. She begins to lurch down the steps, away from the altar, when one of her terrorized sisters suddenly rushes forward. Her determined, wide-eyed expression mirrors the hysterical facial gestures previously exhibited by Sister Cecilia during her direct confrontation with Satan. The older nun, who first discovered Sister Cecilia, shouts to her, but it is too late. The nun's determined look transforms into a lewd facial gesture directed to the old woman, her tongue lasciviously out in mocking defiance, an echo of the Devil's own obscene tongue movements. The contagion of possession is spreading to the entire group. The room explodes in frenzy. Christensen's shot here strongly recalls the earlier scenes of bacchanalian dancing during the Sabbat.

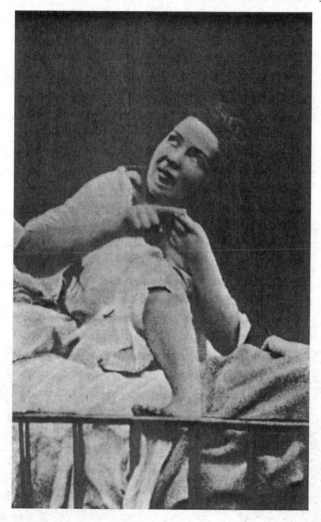

Augustine, "Attitudes passionnelles: Mockery," *Iconographie*, vol. 2, from Désiré-Magloire Bourneville, Paul Régnard, Jean-Martin Charcot, and Édouard Delessert, *Iconographie photographique de la Salpêtrière: Service de M. Charcot*, 3 vols. (Paris: Progrès Médical, 1877–80).

Häxan moves swiftly through a series of images that emphasizes the overwhelming nature of this contagion of possession/ecstasy among the nuns. Close-ups of wild, uncontrolled facial gestures are intercut with long shots of the nuns running riot through the chapel. Satan, watching from his secluded spot behind the altar, flicks his tongue at the women with a look of evil satisfaction. More than at any previous point, the spectacle of hysteria as the root of the witch craze is now visually established in *Häxan*. Christensen makes this explicit when noting that, in a somewhat unconvincingly humanist moment, the women must have suffered greatly "before their nerves abandoned them and insanity broke out." While again referencing as an authoritative source the writings of sisters who suffered from such outbreaks, Christensen does not specifically state whose writings or if he is using a particular case as a model for the scene. His language reflects the analysis of early accounts of religious ecstasy made by Bourneville and others.[7] Specifically, in these scenes he appears to be referring to the famous case of nuns being possessed at Loudun, which was well-known in the literature on witchcraft and possession and has since been eloquently commented on by others, including Michel de Certeau and Michel Foucault.[8] While the film has moved from witchcraft to possession without remark, these scenes nevertheless adhere to the inner logic of the work we have been

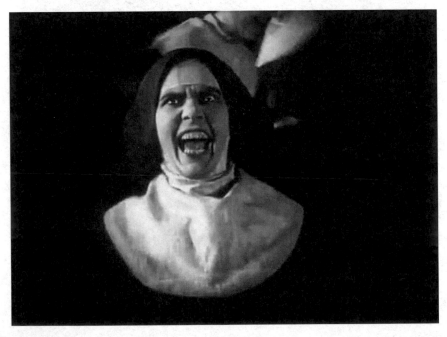

Onset in *Häxan*, film still (Svensk Filmindustri, 1922).

delineating. Though played for spectacular excess, the scandalous acts and grotesquely contorted faces of the nuns correspond directly to the hidden power Christensen is seeking to realize visually in *Häxan*. As Joseph Leo Koerner has observed, "Desecrating the sacred icon, exhibiting it not as object but as abject, they release a strange, transgressive power."[9]

A Mobile Force

Today we can say that the verbal imperialism does not provide the necessary conditions for a real verification. It leaves the possessed few possibilities of resistance, since they themselves "enter" into the system and conform to it. The coding always "works" because the functioning is purely tautological, since the operation takes place within a closed domain. For the exorcists, the difficulty resides not in securing a means of verification of the code, but in keeping "the girls" within the closure of the discourse.

Michel de Certeau, *The Possession at Loudun* (1986)

Häxan has, without announcing it, significantly shifted the focus of the film from witchcraft to "possession." While it is empirically sound to draw a link between the two domains to represent the wider concerns of early modern demonology, they have never been regarded as the same thing.[10] In short, while *Häxan* is still centrally concerned with Satan's relation to humans, particularly women, there is an obvious slippage between the witch and the possessed at this stage. At some level it is a testament to Christensen's power as a filmmaker that this fact is not immediately obvious; indeed, caught by the power of *Häxan*, the viewer can easily pass over the fact that this sequence of the nuns used by the Devil is actually portraying something quite different than a group of women who actively give themselves over to the power of Satan. The ability to make the witch *mobile*, moving beyond the technical, formal definitions of demonologists and modern historians alike, is an effect crucial to the culmination of Christensen's ultimate thesis, which is to link clinical hysteria to these relations with the Devil. Although there remain enormous questions about the cause–effect relation *Häxan* is going for, the strategy of affectively freeing the power of the witch from discursive understandings of witchcraft, possession, or hysteria is crucial—and Christensen executes it effectively. By essentially creating a synecdoche that would credibly fold each discrete entity into the other, Christensen makes obvious the cinematic strategy of the film and foreshadows his attempt to bring the viewer back to science in the final chapter.

The final scene of this chapter makes sense only in the context of the mobility of the witch the film has now established. *Häxan* returns to the chapel, now emptied of possessed nuns save for one young novice. Her face reflects the aftermath of the furiously demonic dance of possession the viewer has just witnessed. Troubled, she kneels before an icon of Christ positioned on the lap of God the Father. The young woman is not praying, however; instead, she slowly reaches up and carefully takes the infant Christ. Cradling the iconographic child, she slowly wanders away. The scene cuts to our familiar, multigenerational trio of inquisitors, working away. Brother John bursts into the room, alerting the older priests that something terrible has occurred. The young woman enters their room, lurching forward, her wild-eyed, frozen face coming into focus. "Burn me at the stake, pious fathers!" she shouts. "Can't you see what the Devil forces me to do?"

The young woman's message conveyed through the intertitle is immediately followed by the image of her spitting on the icon in front of the shocked priests. The image here strongly echoes with the earlier shots of the women abjuring the cross along with other holy instruments at the Sabbat, although this time the blasphemer is forced to do so under obvious duress. The associative link between the two scenes is apparent and not accidental.

Brandishing a cross for protection, Father Henrik's astonishment is precisely what the audience has come to expect. An oddly innovative license is taken in this scene. In empirical terms, the scene does not quite make sense, as there was a clear distinction, both in elite demonological terms and within popular discourses and practices, between an inquisitor and an exorcist. While demonologists would certainly link the two conceptually to the growing power of Satan in the world, it would be imprecise to simply conflate these respective roles. However, both the exorcist and the inquisitor seek the identity of the possessing devil, only in a different order—but in both cases they aim to find the invisible force and name him, and to interview him either directly or indirectly. The distinction was almost always made between those who were the victims of the Devil's malevolence and those who actively worked in concert with this evil force. The nun presenting herself before the inquisitors, begging for help and relief even if this meant her own death, would simply not have been mistaken for a witch.[11]

What would seem like solid empirical ground leads to a narrow interpretation of the scene. There is nevertheless an important point to be considered in what the scene actually does and how it works to set up the final chapter, which brings viewers into present times. "Take me!" she shouts. "Don't you see him? The evil one stands over there and threatens me." The film returns to the woman, her face frozen in terror, reaching out for the one who threatens her. The scene abruptly cuts to black: "To be continued." The power of the witch has been consistently linked to the power of the Devil, yet at this point in *Häxan* it is unambiguously broken free of this seemingly straightforward relation. The power of the witch now exceeds "Satan" as a source and glides beyond the instrumentalized bodies of the witches and the possessed. The last image of this chapter shows a woman reaching out toward this power, toward the camera, toward the audience. *Häxan* breaks off at precisely the moment when the historian, anthropologist,

"Can't you see . . . ?" in *Häxan*, film still (Svensk Filmindustri, 1922).

or film critic would, out of exasperation or desperation, cry foul. Such objectivist hopes aside, we are all now caught by the power of the witch.

At the end of this chapter it is fair to ask what is at stake for Christensen in producing a tension between witchcraft on the one hand and the force of possession on the other. The chapter seems to occupy a sphere of anxiety, where being lost in the crowd, being swept up, and losing all reason at the hands of another are very real possibilities. Through the lens of the postwar period we can appreciate the anxiety produced by the ecstatic possession of the sisters of Loudun as precisely an anxiety about the internalization of authority and its mobile force, an anxiety about a powerful external force taking hold: possession as identification toward destructive ends, as something we might find in Freud's group psychology or Max Weber's writings on authority. While Christensen offers a vague Christian rationality through the invocation of God, it is not always clear what to take from this. In Christensen's account of possession, it is the encounter with evil and the potential for interpellation of its force, to become overpowered, to be taken, to be caught by it, that links possession and witchcraft. And yet, at this moment in the film Christensen is not only expressing his own uneasy relationship with religion, power, and society, he is appealing to the unease of his postwar audience as well.

SEVEN

Hysterias

The history of hysteria seems to cry out for sociological reading.

—MARK S. MICALE, "Hysteria and Its Historiography" (1989)

The most remarkable feature of Benjamin Christensen's presentation of hysteria is what it is not. By 1922, hysteria had been driven to two poles. On the one hand was the image of the "Viennese" hysteric—effete, often affluent, and nearly always female.[1] This figure may be cliché, but it certainly was a real image in the popular imagination of European society in the early twentieth century. Most important, the figure had changed dramatically from the poor, destitute women treated by Pinel and Charcot decades earlier.[2] The female hysteric had evolved.

On the other hand, a different figure of the hysteric had not so much evolved as it was violently forged. He was male, and he was often a soldier, having survived the conflict of the Boer War and the Great War.[3] The sufferer of *kriegsneurose*, war neuroses, or shell shock exists as the living sign of a traumatic event that must be either healed or hidden in order for the society itself to heal. Yet art and cinema take up the challenge in producing experiential works that mine this great catastrophe for whatever purchase

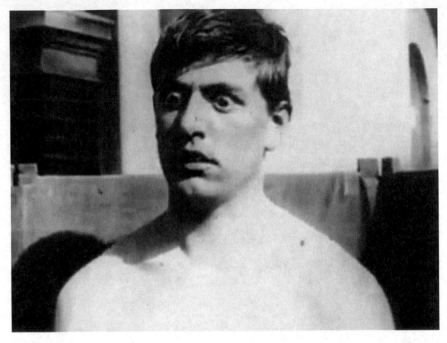

Patient suffering from war neuroses, film still from Arthur Hurst's *War Neuroses: Netley Hospital*, 1917/18). Courtesy of Wellcome Library, London.

may exist for its survivors to truly grasp the event, allowing them to move from traumatically reexperiencing the war to the fading forgetfulness of grieving it. Walter Benjamin is wrong to say that these men returned from the battlefield "silent." While the spirit of his remark is understandable, it is clear that the utterances and gestures of this tragic, unintended iteration simply required a new clinical grammar.[4]

Owing to the fact that Christensen was actively engaged in a project to bring superstition, hypocrisy, and witchcraft into the clear light of modernity, it would seem that this would include a vanguard effort to link his thesis to the changing profile of the hysteric. The configuration of hysteria at the time *Häxan* was made (1918–21) is not one necessarily shared by Christensen's contemporaries; rather, it more closely resembles the hysterics of Charcot's clinic some forty years earlier. Considering that he was working through *the most* technologically modernist art form as a means to illustrate an empirical connection between the witchcraft of *then* and the nervous illness of *now*, it is odd that contemporary "hysterics" fashioned by war only a few years earlier have no real place in the director's account. Yet with its necessary move away from the Freudian centrality of *sexuality*, it is clear that the evolving diagnostics of *war neurosis* would prove to be a serious issue for the respective theses of Freud and Christensen alike.[5]

It should be clear by now that Christensen had not just made a film about either witches or hysterics. Rather, he had made a film about a mobile force that, in particular times and places, has gone by *the name* "witch" or "hysteric." This is the real source of the film's power. It is also a serious formal problem for the director. Having dared to express something of this uncanny power, it becomes impossible for Christensen to close the circle of the thesis from within the framework he has claimed for the film. There is also a fundamental disconnect between the hysteric, whose character is clear, and the witch, who only exhibits the character of the hysteric when under torture. The real "hysterics" were those intent on demonstrating their connection to the Devil without the benefit of torture. All the same, to really follow the expressive logic of the first six chapters of the film to its conclusion would destroy the film's credibility as a documentary work. Thus, Christensen appears to feel that he is faced with a choice that is not really a choice at all. On the one hand, if he seeks to end by attempting to simply rename the witch in the present *Häxan* becomes an exercise of intolerable faux-spirituality. This is the strategy that Margaret Murray took in her own unfortunate attempt to place the witch in history—her transcendentalized earth-goddess version of the figure irritating us to this day through the neutered, self-absorbed identity politics of new-age psychobabble.[6] On the other hand, to fully embrace a position of objectivity would undermine the *idea* of the entire film, the science of hysteria in the 1920s now concerned with the aftermath of a devastating war and the neurotic sway of modernity on delicate psyches rather than the somatic expressions of a mobile power at one time associated with the witch. Christensen is courageous to attempt a conclusion to *Häxan* anyway. He was sufficiently self-aware to know that he could not close this circle, however, as indicated by the fact that he seriously considered dropping this final section entirely when the film was re-released in 1941.

This problem is essential to bear in mind as we confront our own task in analyzing the final chapter of *Häxan*. Considering the explicit depiction of heresy, blasphemy, adjuration, sex, violence, nudity, cannibalism, torture, excretion, perfidy, and deceit that composes the raw materials of *Häxan*'s visual thesis through the first six chapters, it is striking that the most controversial section of the film for contemporary audiences is this final chapter. Christensen's attempt to carry his thesis full circle to issues of mental health, gender, and the law nearly always elicits sustained condemnations from viewers. His articulation of a halfhearted humanism and vague Freudian framework for explaining the witch craze and its relation to contemporary social issues leads many to disregard the conclusion of the film as antiquated and trite. This final chapter serves to further distance *Häxan* from the serious purpose Christensen intended, despite the film's overt tone of studied empirical indifference. While we would agree with the assessment that hysteria does not serve as a sufficient explanatory framework for witchcraft, we obviously cannot agree that *Häxan* is therefore unworthy of serious reflection, for reasons that must already be clear. It is in this context that our analysis of the final chapter (indeed, of the entire film) must be understood.

Satan Dispossessed

This final section of *Häxan* starts with a series of title cards resituating the setting of the film, leaping over "the Devil's possessions," and bringing the audience into the present. Returning to the demonstrative mode that the film began with, Christensen lays out a series of comparisons between the sixteenth and the early twentieth centuries regarding the status and treatment of the stigmatized groups highlighted throughout the film. Starting with old women, *Häxan* notes that such individuals, often alone and lacking independent means of support and care, are now taken in by "pious organizations" and "nursing homes." The scene is visualized through a sequence that takes place in a nursing home, the female residents shown politely eating together around a large table. Christensen shows several of the women, highlighting the effects of aging on their faces and bodies through close-ups. In the course of this sequence, the director draws a visual comparison between the "old hag" of the witch stereotype and the plaintive faces of these present-day elderly women. Despite the focus on physical anomalies (extreme wrinkling, a missing eye, a persistent head shake, a growing "hump" on the back), the gestures of the women suggest kindly old grandmothers rather than the dangerous and repellent figure of the witch. Christensen's visual technique remains consistent with the earlier sections, deploying a nimble editing style in generating flowing close-ups to move the scene along.

Even Maria the "witch" is rehabilitated. Christensen repeats several of the powerful facial close-ups from her earlier interrogation, albeit with a caveat. In keeping with his model of demonstration, the mise-en-scène is punctured as the narrating cards identify the old woman as an actress. Maren Pedersen, rather than her character Maria, is the person we now see. The specter of Maria remains, however, as Christensen tactlessly explains that it would be a mistake to think that beliefs in the Devil are a thing of the past. He tells us that Pedersen once, during a break on the set, stated, "The Devil is real—I have seen him sitting at my bedside." Christensen cuts to a shot of the old woman gazing upward with a gesture of seriousness and trepidation. Although played less "hysterically," the visual rhyme with the final shot of the previous section is unmistakable: a suffering believer catching sight of a devil we (the viewer) cannot ourselves see.

Persisting with this gauche ethnographic display that personalizes the "ignorance" of one of his key collaborators, Christensen begins (with the permission of "the old woman," although it is unclear if he means Maren Pedersen or someone else) an examination of a prayer book, published the year before the film's release, which provides instructions on how to identify the Devil by sight. Flipping through the pages of this slim volume, the helpful offscreen guide returns, his pointer guiding the viewer to the important parts of the pages. The drawings themselves, however, are vague and Christensen does not linger on them. It is unclear at this stage how these contemporary beliefs figure into the director's thesis. Perhaps they are meant to link religious hysteria to the various forms of nervous exhaustion he claims were misidentified as the "insanity" of the witch. So briefly rendered, however, the scene marks the persistence of "superstition" and little more. Christensen appears more eager to move ahead to his "young, nervous woman" than to continue to engage Maren's old wives' tales any further.

A Nervous Young Woman

> The refusal of modern "enlightenment" to treat possession as a hypothesis to
> be spoken of as even possible, in spite of the massive human tradition based on
> concrete experience in its favor, has always seemed to me a curious example of
> the power to fashion in things scientific. That the demon theory will have its
> innings again is to my mind absolutely certain.
>
> William James, *Proceedings of the American Society for Psychical Research* (1909)

At this stage Christensen makes his most direct claim in the film: "The
witch's insanity can be explained as a nervous exhaustion." A title card promises examples to follow, all portrayed by the same actress. Although mim-
icking the dry style of the diagnostician, this almost apologetic statement
hardly qualifies as a diagnosis. In light of the powerful scenes that inten-
sively build to a strong visual association between witchcraft, possession, and
nervous illness that came before, again taking up the arid tone of the scientific demonstration film threatens to disrupt the film's rhythm.

The audience sees the woman. "I have personally known a very nervous
young woman," Christensen assures us. While the specter of the *Malleus
Maleficarum* hangs over the entire film, never is Christensen more explicit
about the *epistemological* analogy that exists between this famous text and
Häxan as he is here. Compare Christensen's attempt to index the film's authority with Institoris and Sprenger's own assertion of expertise, worth citing again, in the *Malleus*:

> We are now labouring at subject matter involving morality, and for this reason
> it is not necessary to dwell on various arguments and explanations everywhere,
> since the topics that will follow in the chapters have been sufficiently discussed
> in the preceding questions. Therefore, we beseech the reader in the name of
> God not to ask for an explanation of all matters, when suitable likelihood is
> sufficient if facts that are generally agreed to be true either on the basis of one's
> own experience from seeing or hearing or on the basis of the accounts given
> by trustworthy witnesses are adduced.[7]

This *demand* for the reader/viewer to take experience, witnessing, and
testimony seriously as empirical evidence is an innovation that was until
recently ignored in historical accounts of demonology and witchcraft.

Christensen's own reliance on the claim that experience and testimony must legitimately occupy the space of evidence is similarly forgotten in the course of the demonstrative cinematic reenactments. Yet both works present themselves as *factual* accounts, whatever our assessment of this claim may be, and work accordingly to innovate and secure their respective definitions of *a fact*.

What has Christensen witnessed? What is the evidentiary value of having "known a very nervous young woman"? First of all, Christensen has witnessed a somnambulist. In turn, we witness the nervous woman serving as Christensen's evidence with a medium shot, the actress in profile, standing upright yet sound asleep. In a posture very similar to that of the possessed nun shown in the previous section abjuring the communion host, Christensen's actress/example extends her arm, palm upright, and turns dreamily toward the camera. The woman's sleepwalking tendency does not factor further in Christensen's argument. Rather, this introductory shot merely serves to introduce the character and index her as a sufferer of the type of nervous disorder that *Häxan* asserts as both cause and explanation for being taken for a witch in earlier times. The somnambulist is a powerful symbol of the melancholic (a "gateway" condition for the witch, as we discussed earlier), images of Charcot's hysteric, and Freud's portrayal of the neurotic in equal measure. The shot requires no further explanation because it brings forth the intersecting categories as a singular image in the mind simply by virtue of what it is. Against the director's better judgment, we have almost come back around to the point where we started, as Christensen finds himself again trying to speak *the language of things*. As before, he will find it impossible to escape *the language of voice* that the inquisitor Visconti centuries before outlined as the proper language of humans.

A Forced Return to the Archive

It may be stated, I believe, as an invariable truth, that, whenever a religion which rests in great measure on a system of terrorism, and which paints in dark and forcible colours the misery of men and the power of evil spirits, is intensely realized, it will engender the belief in witchcraft or magic. The panic which its teachings create, will overbalance the faculties of multitudes. The awful images of evil spirits of superhuman power, and of untiring malignity, will continually haunt the imagination. They will blend with the illusions of age or sorrow or sickness, and will appear with an especial vividness in the more alarming and unexplained phenomena of nature.

W.E.H. Lecky, *History of the Rise and Influence of the Spirit of Rationalism in Europe* (1865)

It is impossible for humans to speak in God's language, the language of things. Thus, a barrage of title cards follows Christensen's display of the somnambulist. The director poses the question of why the woman compulsively reenacts "the very thing she was most afraid she would do." The woman, "like the witch forced by the Devil," compulsively strikes matches, and is shown awake in bed igniting the matches for no apparent reason. Christensen speculates that her compulsion is rooted in the traumatic memory of a fire that once broke out in her home, cutting immediately back to the woman lying uneasily in her bed, eying the box of matches on her nightstand and stiffly reaching out for them. The title cards continue to roll by, informing us that the woman felt she was fighting "an unknown force stronger than her own," followed by a shot where the compulsive grasping for the matchbox is repeated. Christensen's skill as a filmmaker is evident in this sequence, as he has combined rather wordy intertitles with shots of very simple, almost static action in such a way that sense of the unceasing return of the melancholic/hysteric is conveyed to the viewer with an unnoticed economy. And yet it is clear that the director has regressed in his method. Unlike previous chapters of the film, face and tableau on their own no longer move Christensen's thesis forward. He now needs words—an abundance of words—to express what he means.

Christensen also makes a key claim at this stage: the root of the woman's nervous compulsion is one of *excess memory*. This logic would resonate with Freud's theories of the unconscious, of memory, and of trauma for an audience in 1922, and Christensen glosses popular understandings of Freud in a manner that will become all the more obvious when the female character, marked as a hysteric, is visited by an aggressor in her nightmare, repeating a seduction scene. Less obviously but no less crucial to the specificity of *Häxan*'s thesis, the notion of excess memory was also at the heart of theories of melancholia—in the case of Freud, the melancholic's inability to complete the mourning process due to the development of a love attachment to a new object, a chain of (mistaken) associations leading to inevitable fragmentation, which he called melancholy.[8] Furthermore, although limited to an elite group of demonologists and theologians, the mechanics of possession was often linked to concepts akin to excess memory as well, particularly in piecing together how the Devil was able to possess individuals and force them to act according to his will. As Maggi has demonstrated, the logical framework of exorcism not only required techniques that allowed the exorcist to speak the Devil's language himself without becoming possessed but also demanded a renewed agency on the part of the possessed to domesticate and repossess their own memories.[9] In other words, the possessed, like the melancholic and the neurotic, had to be restored to a forgetfulness that Satan simply will not allow of those caught by him.

Christensen emphasizes the connection by again showing us Sister Cecilia in the chapel, driven to unspeakable acts by the Devil. The scene retraces the events shown earlier, although interestingly the sequence is not identical to the one shown before, affording the viewer a slightly altered look at the action. Even this seemingly straightforward strategy of comparing and contrasting case studies across time conveys a sense beyond its explicit purpose in Christensen's hands, as the fluid ontology of memory itself is evoked by presenting the remembered event as a "repetition, but not quite." The director does not explicitly focus on this visual resonance between strategy and content. Rather, he moves directly to his claim that the dazed and delusionary condition of both of these women is characteristic of "nervous diseases we call hysteria." Christensen has finally come around to stating his thesis outright, but the scientific effort to complete his argument is running against the image of the witch he developed in earlier chapters, making it impossible for him to close the circle.

There is still more we are told. Recall the witch who received nightly visitations from Satan in the second chapter. Christensen jumps abruptly to another "memory," this time of the Devil looming over the sleeping couple, the wife writhing responsively as he calls her. As with the previous sequence, the action is the same, but the shots used are slightly different from the ones the audience viewed before. Christensen claims that visitations such as these are still taking place, but in the present it is likely to be "a famous actor, a popular clergyman, or a well-known doctor" who makes the lusty, spectral visit. *Häxan* moves on to a sequence that visually mirrors the earlier visitation depicting the modern hysteric being visited by a well-dressed middle-aged man. Although it is not Satan this time, the appearance of the man is extremely disturbing nevertheless. The actor slowly produces an ominous, creepy smile very similar to the one we see on Johannes's face earlier in the film, and the traumatized woman understandably screams loudly at the sight of him lurking at the foot of her bed. It is here that the understanding of hysteria as generated by a real or imagined seduction scene from which the hysteric continues to suffer (including in traumatic nightmare or apparition scenes, such as this one) comes to recall both Freud's early and abandoned seduction theory and his post-1897 argument that this scene would be repeated in the unconscious. The scene Christensen stages here— so reminiscent of a traumatic nightmare—would not necessarily have to repeat an actual earlier encounter, merely the hysteric's unconscious belief that a consciously undesirable (and unconsciously desired) encounter with a seducer has taken place. Her shriek wakes the others in the house and brings them running to the woman's room; this action is intercut with shots of the leering apparition approaching the terrified woman and placing his hand on her chest. As she rigidly tightens with fear, it remains unclear if he is intending to harm or comfort her. As the others burst into the room, the man dissolves away, much like the Devil would exit the frame. Catatonic, the woman desperately clutches her bed sheets.

The images in this sequence are explicitly modeled on those produced by Charcot, Ronde, Richer, and other pioneers in the diagnosis and treatment of hysteria. As Christensen is quick to claim in the accompanying title card, the woman is displaying the visual characteristics, diagnosed and verified by medical experts, of the modern hysteric.

This is as close to the clinical hysteric as we will get in this final chapter of the film. If we are paying attention, we realize that this is really not very close at all. The demonstrations that surround this visitation scene may display other types of nervous illness such as kleptomania, but these are conditions of a different order. Certainly, theories have been offered that link this set of disorders, but for our purposes here this is unimportant because Christensen does not himself make this claim. At this late stage, the director appears to have rid himself entirely of referencing precise diagnostic categories, a sharp contrast with his approach up to this point. Unlike the meticulous care Christensen took to ground his spectacular images of witches and the Devil in the details of witchcraft's operation in history, here

Transfixed in *Häxan*, film still (Svensk Filmindustri, 1922).

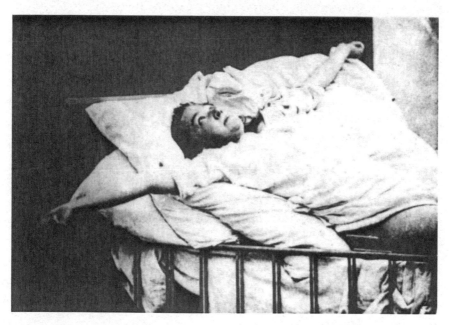

Augustine, "Attitudes passionnelles: Crucifixion," *Iconographie*, vol. 2, from Désiré-Magloire Bourneville, Paul Régnard, Jean-Martin Charcot, and Édouard Delessert, *Iconographie photographique de la Salpêtrière: Service de M. Charcot*, 3 vols. (Paris: Progrès Médical, 1877–80).

we find that the director's shots correspond to the singularity of the hysteric only insofar as he is able to render the *distress of the image* that each figure engenders.[10] Visual spectacle defines *Häxan* as a work; in this final chapter the spectacle threatens to slip from being an instrument of the director to a mere characteristic of his expression.

Christensen *knows* this. He was undoubtedly conscious of the lack we have identified here. It is obvious from the materials he cites for the audience that he is as knowledgeable of the contemporary clinical literature on hysteria as his is of the historical accounts and sources regarding witchcraft. Thus, we are *not* arguing that Christensen could not satisfactorily conclude the film due to an inadequate mastery of his subject. If anything, he intuitively grasped the vector formed out of the convergence of witchcraft, hysteria, and the image so completely that it is these impossible elements that are *really* what is at play in his expression in this final chapter.

Charcot's clinical materials on hysteria, particularly the photographs and the narratives of what came next (as best we can tell) for the models, provide an account that, read against their original intent, yields a similar story to the one we are telling about *Häxan*. As with the witch, this same mobile force that animated the phenomenon refused to be fully encircled or named. The entire spectacular enterprise eventually falls apart, sounding off with what Georges Didi-Huberman has described as the resonating tones of a fugue. Didi-Huberman continues, "Each was asking too much: the physician, with his experimental escalation and his director's vertigo, believing he could do, undo, and redo anything with the bodies yielded to him; the hysteric, with her escalation of consent, in fact demolishing all the reserve and graciousness of representation. What stops there is indeed the reciprocal operation of charm, the death of one desire, if not of two. Disconcertment: deception put out of countenance, the rupture of a rhythm by which a structure could be effusive."[11] The final chapter of *Häxan* does not summarize the director's findings; it expresses his vertigo.

God Forgives Everyone but Satan

We are not quite finished with this sequence in the film. Startlingly, the last shot in this series is of Satan strangling a woman. This seemingly out-of-place edit serves as a relay, moving the narrative forward to consider the correspondence between the Devil's practice of leaving insensitive marks on the bodies of witches and the corresponding symptom of localized loss of sensation common in cases of hysteria—that is, *anesthesia*. Returning to the safety of the distant past, the demonstrative pornography of Christensen's visualization of relations between the Devil and women recurs here. Satan's elongated, clawlike fingers probe a nude woman who is lying with her back to the camera. Although the woman's backside is partially obscured by a strategically placed cauldron, a resurgent sexual overtone dominates the scene. Moving the comparison along, Christensen returns to the executioner checking the Young Maiden's back for these insensitive marks during her interrogation. As expected, there are zones on her back where she cannot discern the executioner's touch. Christensen does not doubt that the loss of sensation is real. The scene cross-fades to a contemporary examination in progress at a doctor's office (interestingly, the doctor is the same leering man who appeared in the nervous woman's room earlier); the viewer is informed that "actually" such insensitivity is a symptom of modern hysteria. As with the comparative mirroring of the night visitations earlier, Christensen has framed this present-day scene as a reflection of its more ancient visual counterpart. The only substantial difference is that the modern woman, nude above the waist, is positioned in a much less salacious manner, muting the sexual overtones of the image and the subsequent affect of obscene desire/repulsion the shots of the Sabbat generate. Again, the literalness of this visual connection weakens a power of correspondence that was at its height in the film when the correlation was suggested according to a subterranean logic that the viewer could easily associate with the witch and the hysteric.

The following scene also functions as a mirror to previous sequences in the film, albeit more loosely given the fact that the correspondence we mark here was likely unintentional. The doctor, having completed his examination of the nervous woman, enters his office, where the afflicted woman's mother is waiting. With all of the oblivious authority the institution confers, the doctor pronounces that, as he suspected, the woman "has hysteria." Again, although less stridently represented, his comportment recalls that of Father Henrik discovering, *as he knew in advance he would*, the witches in his midst. The hysteric, like the witch, will always be found once she is sensed. The option of naming nobody or nothing does not exist at this point—it is solely a question of *whom* at this stage.[12]

Christensen has sought to overtly align his own position with that of science and medicine, at times assuming the casual, total authority of the physician. Bearing this in mind, it is difficult to argue that the correspondence we are drawing here was intentional. The strategy of continually seeking to activate a state in the viewer that provides the ground for a multiplicity of meanings in singular images, however, is not one that can be easily contained according to the subjective intent of the director. Having skillfully cleared this ground, it is impossible for the viewer to move forward and associate the doctor with the inquisitor and the scientist with the demonologist. As such, from our perspective the strength of *Häxan* exceeds even the explicit intent of its director, particularly in scenes such as this one which otherwise resemble the flat clichés that Christensen has methodically sought to empty out. Of course, this power doubles back on the director, aligning him with the inquisitors and doctors. The director has done his best to speak the language of the Devil without himself being possessed; it is clear by the end that this is much more difficult than it seems.

The afflicted woman, still in the examination room adjacent to the office where the conversation between the doctor and the woman's mother is taking place, begins to dress and drifts over to the door to eavesdrop. She overhears the doctor's recommendation that she be detained in the doctor's clinic, lest she have an "unpleasant exchange with the police." She is alarmed by the prospect of being detained in the clinic, her vulnerability emphasized by showing her in a state of near undress. She radiates a sense of susceptibility, of potentially being at the mercy of the doctor or whoever else may walk in on her at any moment.

Häxan then presents a bizarre title card, which is worth reproducing here in full: "Poor little hysterical witch! [Swedish: *Stackars lilla hysteriska häxa!*] In the Middle Ages you were in conflict with the Church. Now it is with the law." Christensen's tone here is extraordinary in that it simultaneously evokes a casual, almost brutal, condescension and at the same time unambiguously claims that women are still being victimized by the various institutions ostensibly in place to protect them. The statement is consistent with Christensen's attempt to align the film with science, but his unthinking paternalism toward the subjects (a paternalism consistent with the very insti-

"Poor little hysterical witch . . ." in *Häxan*, film still (Svensk Filmindustri, 1922).

tutions the film displays) is hard to interpret or ignore. Yes, *Häxan* is a film "of its time," which may account for the particular way in which Christensen expresses himself here. This interpretation is also consistent to some degree with the style of expression evident in the opening chapter of the film. Despite Christensen's skill in producing complex meanings from his images and technique, his ability to allow the viewer to feel the power of the witch even as his overt discourse denies their "real" existence, it is still possible that the exaggerated, even offensive mode of expression at this juncture functions to trigger a more complex response in the viewer. Christensen's willingness to go to extremes calls into question whether even the seemingly careless misogyny of this statement is not, in fact, evocative of a range of associations. The one thing this statement does not do is provide an end to the film.

Stopping Is Not the Same as Ending

> *Häxan* must be watched in the context of logic.
>
> Stan Brakhage, lecture at the Art Institute of Chicago (1973)

How does one bring a film as singular as *Häxan* to an end? It is obvious that Christensen himself never found a satisfactory answer to this question. The final sequence of the nervous woman, caught shoplifting an expensive ring right in front of the male sales clerk in an upscale jewelry store, serves to visualize the "unpleasant exchange" that the doctor predicted for the hysterical woman were she left untreated. The scene is effective on some level in conveying the pathos of the situation, again demonstrating Christensen's skill as a filmmaker. The director's established method of exploiting the expressiveness of the face and its potential for conveying a set of shifting, multiple meanings remains on full display here. The suffering and fear of the afflicted woman is palpable, as is the anger, confusion, and finally the sympathy of the clerk upon catching her in the act. Christensen puts this scene "in present tense" through the woman's reference to the war, the loss of her husband offered as a root cause for her nervous illness. This is the only direct reference to the First World War in *Häxan*.

Christensen does all he can to make us see through what he has conjured with *Häxan*. We are brought back to the "unknown power" in the course of the woman's explanation of her actions, but there is something slightly flat about the point by now. Christensen has to explicitly *say* this in order for the viewer to make the association. This is a moment of defeat for the film. The owner of the store ultimately pardons the woman, but Christensen has created a situation that will not let the director off as easily. Having gone so very far to locate and face the power of the witch, his attempt to "come back," to disavow this power in this final chapter, is no longer possible. Christensen *himself* has made this so. After all of this, Christensen's *Mnemosyne*-like approach has released the power of the images he started with and set them in motion—and once in motion they move of their own accord, ignoring the director's need to call them back in the name of "explanation." Thus, the final montage in the film that attempts to synthesize how far we have come in the centuries since the witch craze fails to put any measurable distance between the contemporary witness and the power of the witch that Christensen has conjured. Intercutting images of the old cosmologies with the new, the director admits that there is not much difference *for him* between the eras after all.

Intuitively knowing that he has gone too far, Christensen returns to forging links across time and space, abruptly claiming that we really do not understand the modern hysteric any more than we understand the witch. Hydrotherapy in the form of a shower, a common treatment for nervous illnesses through the 1920s, is offered as evidence of science's new approach to this phenomenon, but then simultaneously linked visually to the torture chambers of old. It is a *confusing* way to end. The witch will not release her grip on Benjamin Christensen. It is as if he remembers too much, a victim of the excessive memory that he so skillfully illustrated in others. Desperate to draw *Häxan* to an end, the director makes one final cut, inescapably to a bonfire. Women are burning at the stake. Is this *now* or *then*? We reach the darkness at the heart of *Häxan*. We are caught. Christensen wants us to believe that his rational humanist treatment of witchcraft points to a form of hysteria before its time, and yet it is impossible for him to maintain this ostensible objective. The more he pursues his thesis the harder it is for him to stage his thesis cinematically. The tension is between us coming to know the witch according to Christensen, and knowing the witch at all. We are caught, and it is clear that Christensen is caught as well.

It Is Very Hard to Believe . . .

> That mythology is, today, an imaginative exercise for us, should not
> obscure the reality it had for those who lived by it. And since the greater
> part of knowledge of primitive societies was a mythological knowledge,
> the art was an art of knowledge.
>
> —MAYA DEREN, *An Anagram of Ideas on Art, Form, and Film* (1946)

"How would the Devil speak?" This is Benjamin Christensen's question in
a filmed introduction to *Häxan* produced to accompany the 1941 rerelease
of the film. Christensen poses this question as part of his defense of his masterwork as a *silent* film. Hearing voices after the fact would "shatter the
illusion," the director asserts. And yet, as we have continually marked
throughout this book, there is ample evidence that Christensen has a very
good sense indeed of how the Devil speaks. Like the witches, inquisitors,
possessed, hysterics, and doctors that have come before him, the director
displays all the signs of being caught by the mobile power we have largely
been referring to by the name "witch" throughout our own engagement with
the film. This power to touch, to grasp, is almost uniformly disavowed by
more contemporary authorities, but this fact cannot divert our focus from
the signs of being ensnared by a force that resists direct expression or experimental proof. Christensen, nearly twenty years after the fact, betrays himself in the intensity of his relationship with the witch in the short introduction

to his rereleased film. His relish in recounting witch tales, experiential proof culled from his own life and the lives of his friends and acquaintances, overtakes the clinical disposition he starts from in the filmed addendum. The impression left by this strange introduction on the viewer is plain: "It is happening again."

We could ask how Christensen, even two decades after the fact, could truly not *know* that in producing a work as singular as *Häxan* he bore all the signs of being caught by the witch himself—of believing he had objectively, scientifically, humanistically mastered witchcraft while giving the witch the last word. This would be the wrong question to ask, however, as the unmarked tension we are referring to stands as the key to the film's continued influence. If we limit ourselves to judging influence through direct citation of the narrative or the "text," then we must truthfully argue that *Häxan* has had almost no direct influence within the history of cinema. At best, the reformulated Antony Balch version of Christensen's film, titled *Witchcraft through the Ages* (1968), has given the work a persistent, if truncated, presence on the midnight movie circuit. This is a somewhat unfortunate legacy. Balch's strategy of having William S. Burroughs intone curses and commentary in his distinctively flat vocal pitch is a nice addition to this sound version, which is organized according to Burroughs's methodological experiments with cutting up other works.[1] For the most part, however, the tone of this shortened, reformulated version (particularly because of Daniel Humair's wildly inappropriate "free jazz" score) unfortunately tars Christensen's original vision with the brush of being a wacky "head" film.

In the years since the 2001 restoration and limited theatrical rerelease of *Häxan* and its subsequent availability on DVD, critics have retrospectively sought to associate Christensen's film with the emergence of the horror genre, citing its cinematic style as a formal precursor to the foundational classics of this variety that emerged in the 1930s.[2] There is some logic to this claim, but to go any further to argue for *Häxan*'s direct influence on films such as *Dracula* (Tod Browning, 1931) and *Frankenstein* (James Whale, 1931) would be wishful thinking, if for no other reason than Christensen's film went largely unseen at the time of its release and, with the exception of limited rereleases such as the ones we started with here, almost entirely disappeared from the history of the medium until its restoration and reemergence in the first decade of the twenty-first century.

Yet *Häxan* has always been murmuring under the surface of canonical histories of cinema. In order to make sense of this seemingly paradoxical claim we must shift our focus from that of searching for narrative references to the devices the film uses in bringing the power of the witch to life. It is in these devices, sympathetic tools for sensing the *truth* of the witch, where the crucial importance of *Häxan* lies. Within this context we can group the film among those which in the first half of the 1920s called into question the ways cinema could provide an empirical, truthful account of phenomenon located in the "real" world. Alongside films such as *Nanook of the North*, *Häxan*'s heterogeneity in this regard helped to force attempts to formulate genres that could "definitively" separate fact from fiction and gave rise to the Griersonian documentary ideal discussed in Chapter 2. With the benefit of hindsight, we can say that Christensen provided a clear signal as to the impossibility of this ideal, an indication that recurs in later decades through Maya Deren's documentary experiments (*Divine Horsemen: The Living Gods of Haiti*, filmed 1947–48, released 1977), in Jean Rouch's ethno-fictions (*Moi, un Noir*, 1960; *Jaguar*, 1970; *Petit à Petit*, 1971), and via the poetic art–ethnographic meditations of filmmakers such as Robert Gardner (*Dead Birds*, 1963; *Forest of Bliss*, 1985), Harun Farocki (*An Image/Ein Bild*, 1983; *As You See*, 1986; *In Comparison*, 2009), and Lucien Castaing-Taylor (*Sweetgrass*, 2009; *Leviathan*, 2013). Creatively deploying many of the same formal devices that Christensen puts to use in *Häxan*, these filmmakers (among many others) succeed in producing an affective, truthful narrative about the world that rejects dogmatic positions of hard objectivity and absolute relativism alike. Taking their cues as much from artists such as Paul Klee as from empiricists such as John Grierson, all of these films engender a certain grasp of the real through providing an amplified, haptic, *affective* sense of the world. *Häxan* is almost never directly acknowledged as a precursor to nonfiction works such as these; it is our contention that it should be.

This claim brings us back to the viability of Christensen's thesis regarding the relations between the witch and the hysteric. As we have mentioned, while as scholars we cannot wholly accept the director's argument that witchcraft and possession can be fully explained as the result of misrecognized manifestations of hysteria and nervous illness, the logic of the film's thesis, itself derived from the pioneering work of Charcot and Freud, has proven to possess a stronger afterlife than is generally acknowledged. The idea that ecstatic ritual practice, possession, and practices of witchcraft and sorcery bear some relation to modern categories of neurosis and mental illness persisted in the anthropological literature a full fifty years after the release of *Häxan*.[3] In her written work, based on her field studies of Haitian vodun in the late 1940s, Maya Deren both draws an analogy between these seemingly discrete categories of experience and levels a direct criticism at the Western social sciences for the presumed dualisms at work in claiming that ritual ecstasy "really" exists as a form of pathology or simple "cultural difference."[4] Taking Deren's point (largely ignored by anthropologists as being *her* point) further, a long list of studies exist that in some way attempt to link witchcraft, possession, and various elements of psychology without making blunt cause–effect claims in order to elucidate a variety of instances where such occurrences remain an active element of everyday life.[5]

The direct linkages between witchcraft, possession, and hysteria that Christensen asserts have also seen a return in contemporary cinema. Unsurprisingly, given the long-standing tradition represented not only by Christensen and Dreyer but also by a host of filmmakers from Victor Sjöström (*The Phantom Carriage*, 1921) to Ingmar Bergman (*The Seventh Seal*, 1956), of taking up issues of religion, demons, and death, this revived interest in the elements of what is essentially Christensen's thesis has returned via a Nordic filmmaker in Lars von Trier's *Antichrist* (2009). In von Trier's controversial, award-winning film, the female lead ("She") pathologically grieves the loss of her young son, giving rise to "unnatural" visions of talking animals and an aggressively sentient forest surrounding the cabin where her psychotherapist husband ("He") has taken her to effect a "cure." Attributing her own experience variously to "witchcraft" or "gynocide," she is in turn diagnosed by her husband as being a "hysteric," albeit by the end of the film he, too, suffers the visions and afflictions of the unnatural, demonic force that haunts his wife and (perhaps) is to blame for the death of their son. While von Trier does not in any way position *Antichrist* as a nonfiction film, he has explicitly reformulated the devices at work in *Häxan* (down to using the structuring device of "chapters" to organize the narrative) in order to give the viewer a visceral sense of the abject, literary, erotic, mounting power that has variously been associated with the witch, the possessed, and the hysteric. To our knowledge, von Trier has never publicly acknowledged

Häxan or Christensen as a direct influence on *Antichrist*; however, the correspondences between the two films, eighty-seven years after the fact, are plainly obvious.

Return to Malice

So you believe the sciences would have emerged and matured, if they had not been preceded by magicians, alchemists, astrologers, and witches who with their promises and false claims created a thirst, hunger, and taste for hidden and forbidden powers? Indeed, infinitely more has had to be promised than can ever be fulfilled in order that anything at all might be fulfilled in the realm of knowledge.

Friedrich Nietzsche, *The Gay Science* (1882)

In his singularly pitiless style, Nietzsche had identified in 1882 the problem of *promise* raised by a science of man that simultaneously offered a privileged relationship to empirical engagements founded on the cultivated critical acumen of its practitioners to bringing the hidden, unnoticed, invisible secrets of this real into view. Or, as Foucault reminds us, as to their aspiration to the status of science, the human sciences arose out of the search "for the locus of a discourse that would be neither of the order of reduction nor of the order of promise: a discourse whose tension would keep separate the empirical and the transcendental, while being directed at both; a discourse that would make it possible to analyze man as a subject."[6] A remarkable promise . . .

We opened with the silent avowal of this promise in our account of the scholastic debates in the introduction to Part II. What appear to be irreconcilable points of view offered by human scientists such as Malinowski, Rivers, and others were, in the end, genially folded into a discipline (anthropology). The attempt to make man visible through his culture, his society, and his mind cannot help but allow forces other to man to enter zones of observation and experimentation claimed by the human sciences. Christensen, accepting science as epistemology (but hardly bound by science's rules, methods, or common sense) sought a radically different route of access to this hidden force in *Häxan*. Perhaps it is not too much to say that the serene world dreamed by the human scientist of the near past left him unprepared for the invisible forces embedded within such a world.[7] Perhaps it is not stretching things to claim that the social scientist remains just as weak and unprepared when faced with the objects and others of her own dreamworld of the "really real." Benjamin Christensen, for better or worse, turned to face this same hidden power in *Häxan*, an act of turning toward the invisible and the nonsensical. And yet this turning responds to form, or as Dick Houtman and Birgit Meyer suggest, a movement toward materials beyond belief, the material "really real" in the world—or, as Pamela Reynolds offers in her writing on witchcraft, gives clues to the shape of things unseen.[8]

Evidence of Forces Unseen: Some Conclusions

> I think that great art is deeply ordered. Even if within the order there may be enormously instinctive and accidental things, nevertheless I think that they come out of a desire for ordering and for returning fact onto the nervous system in a more violent way.
>
> Francis Bacon, *Interviews with Francis Bacon* (1987)

The image of the witch in *Häxan* is the point of access into her power. This fact constitutes the lasting achievement of Benjamin Christensen's film. In pursuing the cinematic strategy that we have analyzed in detail in this book, Christensen effectively, if imperfectly, addresses himself to the insistent murmuring that would imply that an image cannot of its own accord express something singular about the world contained within it. As difficult as it appears for even the director to accept this at times, *Häxan* stands as a great example of a type of violently naturalist filmmaking that Deleuze associated with directors such as Erich von Stroheim and Luis Buñuel.[9] Again, the key to this claim is the manner in which Christensen tactically takes the viewer down the steepest slope separating the power of the witch from a form of knowing that would deny her. "Down" is the correct directional reference here, as the naturalist line Christensen formulates plunges the viewer directly into the image. We do not simply skirt along the surface of such images, a common accusation leveled at cinematic nonfiction then as now; rather, like the ethnographer and the artist, Christensen's slope is the path by which we descend into the reality of the witch below. Although not referring directly to Christensen, Deleuze's defense of this form of filmmaking as realist applies here as well: "Never has the milieu been described with so much violence or cruelty, with its dual social division 'poor–rich,' 'good men–evildoers.' But what gives their description such force is, indeed, their way of relating the features to an originary world, which rumbles in the depths of all milieux and runs along beneath them."[10]

Häxan also descends into the rumbling depths of an originary world, fathoming the relation between this world and the precarious surface of our own enlightened time. This is a higher form of naturalism, a *perverse* naturalism, which crystalizes in Christensen's film simultaneously as an *ethic* and an *aesthetic*.

The ethics of the film are evident in the fact that Christensen creatively descends into everyday forms of life that are by definition multiple. This fact opposes his work to that which emphasizes a moral position grounded in "identity." It is precisely in Christensen's occasional gestures toward a segmented, identity-based "diversity" where *Häxan* appears to be at times at war with itself. Recognizing that such inconsistencies exist within the work, it is nevertheless clear that Christensen's expression of the witch is most consistent when he poses her power and the myriad attempts to come to grips with this power as a *practical* problem of life rather than an issue of scientific taxonomy or epistemological error. Gaining a grasp of the witch and her other iterations requires another conception of what such forms of life could potentially be, and it is this general disposition that constitutes *Häxan*'s creative importance, anticipating by some two decades the delinking of time from psychological memory and linear causality described by Deleuze as a characteristic of the time-image in cinema after the Second World War.[11] A figure moves differently in relation to multiplicities rather than identity; this accounts for the formal, cinematic power of Christensen's witch.

In short, the success of Benjamin Christensen's *Häxan* is twofold. The ethical force of the work is rooted in its vivid expression of a multiplicitous figure, actively drawn through an unusual, *Mnemosyne*-like procedure that empties and creatively colonizes clichéd figures. The aesthetic innovation lay in its formal strategy to then draw lines among these figures rather than make final points about them. The movement up and down the steep slope between the historical worlds of the image and the originary world that is the source of the witch's unattributable power provides the ground by which Christensen can creatively actualize new forces and powers with which to experiment. This is more than a simple aesthetic: it is a *neuroaesthetic* in that it simultaneously allows for experimentation and diagnosis, pushing the audience toward a phenomenological grasp of the witch or the hysteric that refers to science but exceeds what such rigorously policed forms of knowledge would otherwise teach us.[12] Christensen is certainly not the first to have deployed such a seemingly intolerable method; Duchenne, Muybridge, Marey, and even Charcot himself all exhibited elements of this impossible expression in their own respective works. Going even further in some respects due to his mastery of that most modernist art form, cinema, Christensen realized what he set out to achieve: a truthful engagement with the reality of the witch. Obviously, however, his method and the very object of his inquiry served to complicate things considerably. In expressing the truth of the witch, Christensen could not help but come to be captured by her in the course of this expression.

This is all very hard to believe. Christensen himself never fully comes around to admitting that he does, in fact, believe. And yet, like those figures he depicts in *Häxan*, he *must* believe. The single element that conjoins everything we have described and analyzed in this book demonstrates this and depends on it. Yet this is belief only in the rhetorical sense, and would be more accurately described as a tangle with what Deleuze calls "the powers of the false," making way for an uneasy but equally unambiguous truth—what Luhrmann, in a different context, has described as the epistemological doubling of belief and doubt always integral to the steadfastness of truth and the act of believing.[13] We, too, are caught by Christensen's witch, exhibiting, if not belief as such, then a will to believe in *Häxan*'s world that generates the conditions for an analytic, ethical engagement that would be as familiar to William James as to Gilles Deleuze—to preserve a space for the false to move within the truth of the witch. If nothing else, being caught serves as a testament to what *Häxan* can do.

ACKNOWLEDGMENTS

While writing this book we incurred many debts—some individual, some institutional. Stefanos Geroulanos has played many roles throughout: advocate, interlocutor, editor, diplomat, critic, and most of all, fierce friend. Henning Schmidgen, Stuart McLean, and Walter Stephens shared their insights and expertise on the penultimate version of the book, and we benefited greatly from their generosity. Tom Lay and the staff at Fordham University Press have been careful and skillful in their work to see the book to publication. Our conversations about *Häxan* began with Helen Tartar. Her enthusiasm was as infectious as Christensen's witch, and she is missed. We were fortunate enough to spend time writing, thinking, and walking at the Cove Park Arts Centre in Kilcreggan, Scotland, thanks to the generous invitation of its director, Julian Forrester. Two Americans finishing a book about a Danish filmmaker's Swedish film along the expanse of a Scottish loch seems somehow apt. We are very thankful as well to many others who have been charitable in their support, offering insightful comments during presentations of chapters in early stages of development at the Museum of the University of St. Andrews and the Humanities Center at Wayne State University. Wayne State University and the School of Social and Political Science, University of Edinburgh provided financial support for our research and writing, for which we are grateful. During the period of revision, one of us (Todd) was Residency Research Fellow at the Eisenberg Institute for Historical Studies at the University of Michigan, whose faculty and students proved a tremendous intellectual resource. We would also like to thank Suzan Alteri at the Wayne State University Purdy–Kresge Library for research assistance and Patricia Allerston at the National Galleries of

Scotland who, as the internal curator of the exhibition *Witches and Wicked Bodies* in 2013, has been a great help in obtaining permissions for several of the images in this book and very generously allowed us to present some of this material as part of the public educational series for the exhibition. Richard would also like to thank his friends at The Cube and at Simpson Loan in Edinburgh for providing a friendly place to work and endless coffee to power it along. Finally, special thanks go to Magnus Course, who first drew our attention to the similarities between *Häxan* and Lars von Trier's *Antichrist*, and to Lotte Hoek for her steady support of our project over the years and her valuable advice along the way. Our preparation of the bibliography of Christensen's source materials owes much to film scholar Casper Tybjerg's "*Bibliothèque diabolique*: A Photographic Exploration of Christensen's Historical Sources," found on the extra features of the Criterion Collection rerelease of the film in 2001.

Elements of the introductory section to Part I have previously appeared in an essay authored by one of us (Richard) in a different form as "Knowing Primitives, Witches, and the Spirits: Anthropology and the Mastery of Nonsense," *Republic of Letters: A Journal for the Study of Knowledge, Politics, and the Arts* 4, no. 1 (2013): 1–22.

INTRODUCTION: WHAT IS *HÄXAN?*

1. Casper Tybjerg, "Images of the Master," in *Benjamin Christensen: An International Dane*, ed. Jytte Jensen (New York: Museum of Modern Art, 1999), 8–22.

2. Benjamin Christensen, "The Future of Film," in *Benjamin Christensen: An International Dane*, ed. Jytte Jensen (New York: Museum of Modern Art, 1999), 38–39.

3. Tybjerg, "Images of the Master." See also Jack Stevenson, *Witchcraft through the Ages: The Story of "Häxan," the World's Strangest Film and the Man Who Made It* (Goldaming, U.K.: FAB Press, 2006).

4. As quoted by Arne Lunde in "Benjamin Christensen in Hollywood," in *Benjamin Christensen: An International Dane*, ed. Jytte Jensen (New York: Museum of Modern Art, 1999), 23.

5. These films were *The Devil's Circus* (1926), *Mockery* (1927), and *The Mysterious Island* (1929; completed by Lucien Hubbard).

6. Christensen's departure was not noted in the Hollywood trade press, and he had been in Copenhagen for fully one year before the Danish press learned of his return. Arne Lunde, "The Danish Sound Feature at Nordisk," in *Benjamin Christensen: An International Dane*, ed. Jytte Jensen (New York: Museum of Modern Art, 1999), 34.

7. These films were *Children of Divorce* (*Skilsmissens Børn*, 1939), *The Child* (*Barnet*, 1940), and *Come Home with Me* (*Gaa med mig hjem*, 1941).

8. Benjamin Christensen, "Benjamin Christensen's film: En studie over häxprocesserna," *Filmjournalen* 21–22 (December 25, 1921): 737, quoted in Tybjerg, "Images of the Master," 15.

9. James Siegel, *Naming the Witch* (Stanford, Calif.: Stanford University Press, 2006), 21–25.

10. Jeanne Favret-Saada, *Deadly Words: Witchcraft in the Bocage* (Cambridge, U.K.: Cambridge University Press, 1980), 4.

11. Ibid., 17–18.

12. For examples of this phenomenon within what are otherwise divergent arguments see Favret-Saada, *Deadly Words*; and Siegel, *Naming the Witch*. See also Claude Lévi-Strauss, *Structural Anthropology* (New York: Basic Books, 1963); and Peter Geschiere, *The Modernity of Witchcraft: Politics and the Occult in Postcolonial Africa*, trans. Janet Roitman and Peter Geschiere (Charlottesville: University Press of Virginia, 1997), and *Witchcraft, Intimacy, and Trust: Africa in Comparison* (Chicago: University of Chicago Press, 2013).

13. Steven C. Caton, *"Lawrence of Arabia": A Film's Anthropology* (Berkeley: University of California Press, 1999).

14. Georges Didi-Huberman, *Invention of Hysteria: Charcot and the Photographic Iconography of the Salpêtrière*, trans. Alisa Hartz (Cambridge, Mass.: MIT Press, 2003).

15. The full title of this release is *Häxan: Witchcraft through the Ages*, which combines Christensen's original Swedish name for the work with the title that was often used when the film was shown in English-speaking countries. For the purposes of this book we have used the restored version of *Häxan* that was released commercially on DVD by the Criterion Collection in 2001. While other versions of the film exist, the differences that one finds in these alternate versions are immaterial to our argument in this book. For example, the print of *Häxan* that was screened theatrically in November 2013 as part of an event jointly sponsored by the British Film Institute (BFI) and National Museums Scotland (NMS) as part of the NMS exhibition *Witches and Wicked Bodies* in Edinburgh differs from the Criterion version in several respects. In particular, the film's first chapter in the BFI print is truncated, several sequences in Chapter 4 are edited "out of order" (likely the result of restoration undertaken after the censors snipped away at the original print), and the frame tinting is at times different from what one finds in the Criterion version.

16. John Ernst, *Benjamin Christensen* (Copenhagen: Danske Filmmuseum, 1967); Jytte Jensen, "Benjamin Christensen: An International Dane," in *Benjamin Christensen: An International Dane*, edited by Jytte Jensen (New York: Museum of Modern Art, 1999), 4–6; Tybjerg, "Images of the Master"; Tybjerg, "Red Satan: Carl Theodor Dreyer and the Bolshevik Threat," in *Nordic Explorations: Film before 1930*, ed. John Fullerton and Jan Olsson (Sydney: John Libbey, 1999), 19–40; Tybjerg, audio commentary, *Häxan: Witchcraft through the Ages* (New York: Criterion Collection, 2001); Arne Lunde, "Scandinavian Auteur as Chameleon: How Benjamin Christensen Reinvented Himself in Hollywood, 1925–29," *Journal of Scandinavian Cinema* 1, no. 1 (2010): 7–23.

17. The Danish psychologist Alfred Lehmann is another a resource for Christensen on the relationship of psychiatry and witchcraft. See Alfred Lehmann, *Hypnosen og de dermed beslaegtede normale Tilstande* (Copenhagen:

Philipsen, 1890), and *Overtro og trolddom fra de aeldste tider til vore dage* (Copenhagen: J. Frimodts Forlag, 1893–96); as well as the German psychiatrist Otto Snell, *Hexenprozesse und Geistesstörung* (Munich: J. F. Lehmann, 1891).

THE WITCH IN THE HUMAN SCIENCES AND THE MASTERY OF NONSENSE

1. We are not the first to remark on the concern with invisible objects and forces in the process of ascertaining what counts as evidence. Georges Didi-Huberman, defending the contemporary evidentiary value of photographs taken by Sonderkommando prisoners in Auschwitz that directly depict the extermination of Jews (and thus undercutting the widely held view of the Holocaust as fundamentally "unrepresentable"), writes, "The entire history of images can thus be told as an effort to *visually transcend* the trivial contrasts between the *visible* and *invisible*." *Images in Spite of All: Four Photographs from Auschwitz*, trans. Shane B. Lillis (Chicago: University of Chicago Press, 2008 [2003]), 133. Further, Bruno Latour writes that "invisibility in science is even more striking than in religion: hence nothing is more absurd than the opposition between the visible world of science and the 'invisible' world of religion." *On the Modern Cult of the Factish Gods* (Durham, N.C.: Duke University Press, 2010), 93. In Latour's case, a reoccurring theme in his otherwise wide-ranging oeuvre is with the development of a theory of images that accounts for the invisible and for the processes by which the unseen is brought into view as evidence (see especially Latour's lesser-known anthropological research in the Ivory Coast and his studies in biblical exegesis as well as his now famous work on the transmission of knowledge and experience in science and art). For a comprehensive critical review of Latour's intellectual career, see Henning Schmidgen, *Bruno Latour in Pieces: An Intellectual Biography*, trans. Gloria Custance (New York: Fordham University Press, 2015 [2011]). Taking things a step further, Christian Suhr and Rane Willerslev emphasize that in documentary and ethnographic film it is the formal technique of *montage* that allows the medium to convey "an invisible and irreducible otherness" that would otherwise be unavailable to the scientist and to everyday people alike. "Can Film Show the Invisible? The Work of Montage in Ethnographic Filmmaking," *Current Anthropology* 53, no. 3 (2012): 282.

2. Jonathan Strauss, *Human Remains: Medicine, Death, and Desire in Nineteenth-Century Paris* (New York: Fordham University Press, 2012), 46.

3. There is an important technical divergence that takes place in the nineteenth century that would distinguish respective claims in the sciences such as medicine and biology and change their ability to make appear through imaging and sensing technologies realms of the world previously held to be beyond empirical proof. Able to argue for a world "below the threshold of

perception," medicine, biology, physics, and other sciences were able to define their relation to the nonsensical via a "visibility to come" embodied in then-new technologies such as photography and advances in older technologies such as the microscope and the telescope. While reaching toward new points on an ever-widening horizon, mastery over the nonsensical here holds out the literal promise of revealing the heretofore unseen. Imaging technologies act as the correction of a deficiency, this (essential) deficiency being invisibility or insensibility itself. See Corey Keller, "Sight Unseen: Picturing the Invisible," in *Brought to Light: Photography and the Invisible, 1840–1900*, ed. Corey Keller (San Francisco: San Francisco Museum of Modern Art, 2008), 20–21.

4. Lorraine Daston and Katharine Park, *Wonders and the Order of Nature, 1150–1750* (New York: Zone Books, 1998), 221–23. This passage also obviously echoes Michel Foucault's core argument regarding the constitution of "man" as an empirical object in the formation of the human sciences of the nineteenth century in *The Order of Things: An Archaeology of the Human Sciences* (New York: Vintage 1970 [1966]).

5. We mean "dark precursor" here in the manner that Gilles Deleuze has used the term. Specifically, when writing of communication between heterogeneous systems, Deleuze writes, "here we must pay the greatest attention to the respective roles of difference, resemblance, and identity. To begin with, what is this agent, this force which ensures communication? Thunderbolts explode between different intensities, but they are preceded by an invisible, imperceptible *dark precursor*, which determines their path in advance but in reverse, as though intagliated. Likewise, every system contains its dark precursor which ensures the communication of peripheral series. As we shall see, given the variety among systems, this role is fulfilled by quiet diverse determinations. The question is to know in any given case how the precursor fills this role." *Difference and Repetition*, trans. Paul Patton (New York: Columbia University Press, 1994 [1968]), 119. We can think of no more precise a description of the aim of this book than the one Deleuze provides here.

6. George W. Stocking, *The Ethnographer's Magic, and Other Essays in the History of Anthropology* (Madison: University of Wisconsin Press, 1992), 51.

7. Bronislaw Malinowski, *Argonauts of the Western Pacific: An Account of Native Enterprise and Adventure in the Archipelagoes of Melanesian New Guinea* (London: Routledge 2002 [1922]), 19.

8. George W. Stocking and James Clifford have individually offered critical summaries of these critiques. See George W. Stocking, "The Ethnographer's Magic: Fieldwork in British Social Anthropology from Tylor to Malinowski," in *Observers Observed: Essays on Ethnographic Fieldwork*, ed. George W. Stocking, vol. 1 of *History of Anthropology* (Madison: University of

Wisconsin Press, 1983), 70–120. See also James Clifford, "On Ethnographic Authority," *Representations* 1, no. 1 (1983): 118–46.

9. While Malinowski consistently imparts an optimistic, humanist sheen to the trial of fieldwork, the mythical undertones of such research as the only proper "rite of passage" for the anthropologist are quite evident in the scholastic work he published during his lifetime. Even in light of such clues, the crude shock of Malinowski's personal struggles with loneliness, racist antipathy for his interlocutors, and his feelings of ambivalence and even persecution as factors the fieldworker must suffer and overcome for the sake of scientific knowledge of the Other created a momentary crisis within the discipline when they were brought to light via the posthumous publication of his fieldwork diary. While the racism, desire, and frank neurosis that Malinowski displays in the diary were rightfully the subject of a great deal of discussion, the fact that the account takes the form of a personal spiritual trial to be overcome has gone largely unnoticed. See Bronislaw Malinowski, *A Diary in the Strict Sense of the Term* (London: Routledge and Kegan Paul, 1967).

10. Dan Sperber, *On Anthropological Knowledge* (Cambridge, U.K.: Cambridge University Press, 1982), 54–60.

11. Joseph Leo Koerner, *The Reformation of the Image* (Chicago: University of Chicago Press, 2004), 206, 210.

12. Norman Cohn, *Europe's Inner Demons: An Enquiry Inspired by the Great Witch-Hunt* (London: Chatto Heinemann, 1975).

13. There is obviously a vast literature on the *Malleus*. For a reading of the reactions against the *Malleus* see Kors and Peters, "Scepticism, Doubt, and Disbelief in the Sixteenth and Seventeenth Centuries" (introduction to Part 7), in *Witchcraft in Europe, 1100–1700: A Documentary History* (Philadelphia: University of Pennsylvania Press, 1972), 311–13. Also, see the discussion of the *Malleus Maleficarum* in relation to other demonological texts, and the crisis of doubt that each produced, in Hans Peter Broedel, *The "Malleus Maleficarum" and the Construction of Witchcraft: Theology and Popular Belief* (Manchester, U.K.: Manchester University Press, 2003); Thomas Willard Robisheaux, *The Last Witch of Langenburg: Murder in a German Village* (New York: Norton, 2009).

14. Christopher S. Mackay, *The Hammer of Witches: A Complete Translation of the "Malleus Maleficarum"* (Cambridge, U.K.: Cambridge University Press, 2009 [1487]), 261. For a different translation of the same passage see Broedel, *"Malleus Maleficarum" and the Construction of Witchcraft*," 20–21.

15. Quoted in Stuart Clark, *Thinking with Demons: The Idea of Witchcraft in Early Modern Europe* (Oxford, U.K.: Oxford University Press, 1997), 135.

16. For starkly different accounts of medieval millenarian heretics compare Norman Cohn, *The Pursuit of the Millennium: Revolutionary Millenarians*

and Mystical Anarchists of the Middle Ages (London: Pimlico, 2004 [1970]); with Raoul Vaneigem, *The Movement of the Free Spirit* (New York: Zone Books, 1994 [1986]).

17. Armando Maggi, *Satan's Rhetoric: A Study in Renaissance Demonology* (Chicago: University of Chicago Press, 2001), 14.

18. St. Thomas Aquinas, *Summa theologiae*, ed. Thomas Gilby et al., 60 vols. (New York: McGraw Hill, 1964), I.2.3 [2:12–13].

19. These proofs, as articulated in *Summa theologiae* are as follows: the argument of the unmoved mover; the argument of the first cause; the argument from contingency; the argument from degree; and lastly, the teleological argument.

20. Jeffrey Burton Russell, *Witchcraft in the Middle Ages* (Ithaca, N.Y.: Cornell University Press, 1972), 10–14.

21. Walter Stephens, *Demon Lovers: Witchcraft, Sex, and the Crisis of Belief* (Chicago: University of Chicago Press, 2002), 202.

22. Reflecting the growing importance of Satan as a core element in proofs of God's existence, the characteristics of the Devil become correspondingly elaborated in scholastic treatments of the issue from roughly the year 1050 onward. Jeffrey Burton Russell's analysis of the increasingly extravagant descriptions of Satan offered by figures such as Anselm of Canterbury, Peter Lombard, and Thomas Aquinas aptly demonstrates that, lacking the capacity to empirically describe God, theologians took to offering detailed descriptions of Satan instead as a kind of secondary representational strategy. See Chapter 7, "The Devil and the Scholars," in Jeffrey Burton Russell, *Lucifer: The Devil in the Middle Ages* (Ithaca, N.Y.: Cornell University Press, 1986), 159–207.

23. The idea that demons were the first to recognize Christ's divinity is well established in the New Testament. For example, an account of Jesus exorcising a possessed man in a temple appears in both the gospels of St. Mark and St. Luke: "And there was in their synagogue a man with an unclean spirit; and he cried out, saying, 'Let us alone: what have we to do with thee, thou Jesus of Nazareth? Art thou come to destroy us? I know thee who thou art, the Holy One of God.' And Jesus rebuked him, saying, 'Hold thy peace, and come out of him.' And when the unclean spirit had torn him, and cried with a loud voice, he came out of him. And they [the disciples] were amazed, insomuch that they questioned among themselves, saying 'What thing is this? What new doctrine *is* this?' For with authority commandeth he even the unclean spirits, and they do obey him. And immediately his fame spread abroad throughout all the region round about Galilee." Mark 1: 23–28. This passage appears almost verbatim in Luke 5: 33–37. For a contemporary historical account of how these biblical accounts presented a series of delicate epistemological problems for sixteenth-century theologians and clerics, see

Chapter 2, "Possessed Behaviors," of Nancy Caciola's *Discerning Spirits: Divine and Demonic Possession in the Middle Ages* (Ithaca, N.Y.: Cornell University Press, 2003), 31–78.

24. Broedel, *"Malleus Maleficarum" and the Construction of Witchcraft*," specifically his chapter "Misfortune, Witchcraft, and the Will of God," 66–90.

25. Stephens, *Demon Lovers*, 167.

26. *Malleus Maleficarum*, Part III, Q. 15.

27. The status of witnessing as the production of truth comes in to its own in the confession, as surely as it did roughly a century later in Robert Boyle's *New Experiments* (1660), which revolutionized practical experimental procedures in the laboratory for generations to come. For an excellent account of how testimony came to be admissible as evidence in laboratory settings via Boyle's *New Experiments*, see the "Prototype .08" section of Part 2, "Trial Runs," in Avital Ronell's *The Test Drive* (Urbana: University of Illinois Press, 2005), 88–102.

28. Marcel Griaule, *Methode de l'etnographie* (Paris: Presses Universitaires de France, 1957), 59. Our thanks to Stefanos Geroulanos for pointing out this passage to us and for providing us with his translation, which is cited here. See also James Clifford, *The Predicament of Culture: Twentieth-Century Ethnography, Literature, and Art* (Cambridge, Mass.: Harvard University Press, 1988), 68, 71–76, 83–84.

29. Griaule, *Methode de l'etnographie*.

30. Despite Griaule's well-known political opposition to Michel Leiris and Claude Lévi-Strauss, it is worth noting that he was hardly unique in these practices. Lévi-Strauss, who cast the anthropologist (himself) relentlessly as the guilty conscience of modern, colonial European states, reports in *Tristes tropiques* having used quite the same techniques as Griaule—dissimulation, extraction of secret information, and the assumption that one should not take statements made by interlocutors at face value. While expressed in a more sympathetic language than Griaule typically deployed, Lévi-Strauss's own suspicions of "natives" is evident in his accounts of attempting to buy Kaingang utensils which were "owned" by a four-year-old-girl incapable of transacting a negotiated sale (158) and the "inability" of Caduveo artists to "remember" how to reproduce complex motifs in their body painting practices (187) for the benefit of the ethnographer. He goes further when, seeking to learn the "real" names of his Nambikwara interlocutors (and thus explicitly violate their taboo on the open use of proper names), Lévi-Strauss takes advantage of what begins as a linguistic misunderstanding with a little girl to trick some children into revealing first their own "real" names and eventually the names of the adults as well. Lévi-Strauss dryly admits that this was "rather unscrupulous" on his part (279). In *Tristes tropiques*, trans. John Weightman and Doreen Weightman (New York: Penguin, 1992 [1955]).

31. Ronell, *Test Drive*, 110.

32. Jacques Derrida, *Demeure: Fiction and Testimony*, trans. Elizabeth Rottenberg (Stanford, Calif.: Stanford University Press, 2000), 30. See also Ronell, *Test Drive*, 110–11.

33. Clifford Geertz, *Works and Lives: The Anthropologist as Author* (Stanford, Calif.: Stanford University Press, 1988), 77. See also Anna Grimshaw, *The Ethnographer's Eye: Ways of Seeing in Modern Anthropology* (Cambridge, U.K.: Cambridge University Press, 2001), 51–56.

34. Franz Heinemann, *Der Richter und die Rechtspflege in der deutschen Vergangeheit: Mit 159 Abbildungen und Beilagen nach den Originalen aus dem fünfzehnten bis achtzehnten Jahrhundert* (Leipzig: Eugen Diederichs, 1900).

35. Eduard Fuchs, *Illustrierte Sittengeschichte vom Mittelalter bis zur Gegenwart* (Munich: Langen, 1912).

36. Julie Epstein, *Altered Conditions: Disease, Medicine, and Storytelling* (New York: Routledge, 1995), 26, 35–36; Philippe Huneman, "Writing the Case: Pinel as Psychiatrist," *Republics of Letters: A Journal for the Study of Knowledge, Politics, and the Arts*, 3, no. 2 (2014): 1–28; in the case of Freud, specifically his Dora (1904), Rat-Man (1909), and Wolf-Man (1917). See Sigmund Freud, *Dora: An Analysis of a Case of Hysteria* (New York: Touchstone, 1997 [1905]), and *Three Case Histories: The "Wolf Man," the "Rat Man," and the Psychotic Doctor Schreber* (New York: Touchstone, 1996).

37. Peter Galassi, *Before Photography: Painting and the Invention of Photography* (New York: Museum of Modern Art, 1981), 25, 29.

38. A number of excellent studies exist regarding the relation between science and image making during the late nineteenth century, including Marta Braun, *Picturing Time: The Work of Etienne-Jules Marey (1830–1904)* (Chicago: University of Chicago Press, 1995); François Dagognet, *Etienne-Jules Marey: A Passion for the Trace*, trans. Robert Galeta with Jeanine Herman (New York: Zone Books, 1992); François Delaporte, *Anatomy of the Passions*, trans. Susan Emanuel and ed. Todd Meyers (Stanford, Calif.: Stanford University Press, 2008); Georges Didi-Huberman, *Invention of Hysteria: Charcot and the Photographic Iconography of the Salpêtrière*, trans. Alisa Hartz (Cambridge, Mass.: MIT Press, 2003); Phillip Prodger, *Time Stands Still: Muybridge and the Instantaneous Photography Movement* (Oxford, U.K.: Oxford University Press, 2003); and Rebecca Solnit, *River of Shadows: Eadweard Muybridge and the Technological Wild West* (New York: Viking, 2003).

39. Jan Goldstein's work regarding Nanette Leroux, who was one of the first to be diagnosed with "hysteria" in the modern sense (1820s), is a good source for the detailed outlines of the long period where possessions were increasingly understood in medicalized terms. See Jan Goldstein, *Hysteria*

Complicated by Ecstasy: The Case of Nanette Leroux (Princeton, N.J.: Princeton University Press, 2010).

40. For an excellent summary of why Charcot would frame the relation in this way, see the introduction in Caciola's *Discerning Spirits*, 1–30.

41. Sister Jeanne des Anges, *Soeur Jeanne des Anges, supérieure des Ursulines de Loudun, XVIIe siècle: Autobiographie d'une hystérique possédée, d'après le manuscrit inédit de la bibliothèque de Tours (Bibliothèque diabolique)* (Paris: Progrès Médical, 1886).

42. Désiré Magloire Bourneville, *Science et miracle: Louise Lateau ou la stigmatisée belge (Bibliothèque diabolique)* (Paris: Progrès Médical, 1875).

43. François Buisseret and Désiré Magloire Bourneville, *La possession de Jeanne Fery, religieuse professe du Couvent des Soeurs noires de la ville de Mons, 1584 (Bibliothèque diabolique)* (Paris: Progrès Médical, 1886).

44. Bourneville, *Science et miracle*, i.

45. Caciola characterizes exorcists in the Middle Ages in basically the same way in *Discerning Spirits*, 2.

46. Jean-Martin Charcot and Paul Marie Louis Pierre Richer, *Les démoniaques dans l'art: Avec 67 figures intercalées dans le texte* (Paris: Delahaye et Lecrosnier, 1887).

47. Georges Gilles de la Tourette, *Traité clinique et thérapeutique de l'hystérie d'après l'enseignement de la Salpêtrière* (Paris: E. Plon, Nourrit et Cie, 1891).

48. Paul Auguste Sollier, *Genèse et nature de l'hystérie, recherches cliniques et expérimentales de psycho-physiologie* (Paris: Félix Alcan, 1897).

49. A. R. G. Owen, *Hysteria, Hypnosis and Healing: The Work of J.-M. Charcot* (London, Dobson, 1971).

50. Ibid.

51. *Hippocratic Writings*, trans. J. Chadwick and W. N. Mann (London: Penguin, 1983), 237–51.

52. The evolving conceptualization of hysteria between Hippocrates and the nineteenth century is not uncomplicated, as Jan Goldstein makes clear; see Goldstein, *Hysteria Complicated by Ecstasy:*, 51–55. Also see Goldstein, *Console and Classify: The French Psychiatric Profession in the Nineteenth Century* (New York: Cambridge University Press, 1990); and Régine Plas, *Naissance d'une science humaine: La psychologie* (Paris: Presses Universitaires de Rennes, 2000).

53. Michel Foucault, *Madness and Civilization: A History of Insanity in the Age of Reason*, trans. Richard Howard (New York: Vintage, 1965), 38–40.

54. Caciola, *Discerning Spirits*, 86.

55. Ulrich Baer, *Spectral Evidence: The Photography of Trauma* (Cambridge, Mass.: MIT Press, 2002), 31.

56. Josef Breuer and Sigmund Freud, *Studies on Hysteria* (London: Penguin, 1991 [1895]); Sigmund Freud, *Dora: An Analysis of a Case of Hysteria* (New York: Touchstone, 1997 [1905]).

57. Traugott K. Oesterreich, *Occultism and Modern Science* (London, Methuen, 1923).

58. Pierre Janet, *Névroses et idées fixes* (Paris: Félix Alcan, 1898), 32.

59. Ibid.

60. Oesterreich, *Occultism and Modern Science*, 23.

I. EVIDENCE, FIRST MOVEMENT: WORDS AND THINGS

1. See Armando Maggi's *Satan's Rhetoric: A Study of Renaissance Demonology* (Chicago: University of Chicago Press, 2001), 109: "As Zacharia Visconti explains at the beginning of *Complementum artis exorcisticae*, there are three kinds of words: 'Threefold is language. The first kind is the language of deed; the second of the voice, and the third is of the mind. God speaks with the language of deed. Human beings speak with the language of the voice, and angels speak with the language of the mind.'" Visconti notes that a common sign of demonic possession is the loss of voice, and that the goal of exorcism is "to impose silence on the devil."

2. Ibid., 180–224.

3. Catherine Russell, "Surrealist Ethnography: *Las Hurdes* and the Documentary Unconscious," in *F Is for Phony: Fake Documentary and Truth's Undoing,* ed. Alexandra Juhasz and Jesse Lerner (Minneapolis: University of Minnesota Press, 2006), 99–115. See also Alison Tara Walker, "The Sound of Silents: Aurality and Medievalism in Benjamin Christensen's *Häxan,*" in *Mass Market Medieval: The Middle Ages in Popular Culture,* ed. David W. Marshall (Jefferson, N.C.: McFarland, 2007), 42–56.

4. Gaston Maspero, *Egyptian Archaeology* (New York: G. P. Putnam's Sons, 1887). See also Stuart Clark, *Thinking with Demons: The Idea of Witchcraft in Early Modern Europe* (Oxford, U.K.: Oxford University Press, 1997).

5. Richard Kieckhefer, *European Witch Trials: Their Foundations in Popular and Learned Culture, 1300–1500* (London: Routledge, 1976), and *Magic in the Middle Ages* (Cambridge, U.K.: Cambridge University Press, 1989).

6. Charles Zika, *Exorcising Our Demons: Magic, Witchcraft and Visual Culture in Early Modern Europe* (Leiden: Brill, 2003), and *The Appearance of Witchcraft: Print and Visual Culture in Sixteenth-Century Europe* (London: Routledge, 2008).

7. Edward Burnett Tylor, *Primitive Culture: Researches into the Development of Mythology, Philosophy, Religion, Language, Art, and Custom,* 2 vols. (London: John Murray, 1871).

8. Friedrich Max Müller, *Lectures on the Science of Religion* (New York: Charles Scribner and Company, 1872).

9. George Rawlinson, *History of Ancient Egypt* (London: Longman and Green, 1881).

10. Maspero, *Egyptian Archaeology*.

11. Hartmann Schedel, *Liber Chronicarum* (Augsburg, 1497).

12. Frances Yates, *The Occult Philosophy in the Elizabethan Age* (London: Routledge, 2001 [1979]), 1.

13. Frances Yates, *Giordano Bruno and the Hermetic Tradition* (London: Routledge, 2002 [1964]), 2.

14. Tommaso Campanella, *The City of the Sun: A Poetical Dialogue* (Berkeley: University of California Press, 1981 [1623]). For a full account of the central importance of architecture, place, and the cultivation of the art of memory and the self in Renaissance Hermeticism, see Frances Yates, *The Art of Memory* (London: Pimilco, 1992 [1966]).

15. Ficino published these collected writings under the title *Pimander*.

16. It is true the writings that fueled the radical thought of Agrippa and Bruno were not based on the ancient Egyptian wisdom of Hermes Trimesgistus, but rather on anonymous Greek authors writing between 100 and 300. Yates, *Giordano Bruno*, 2–3.

17. For an excellent account of how Renaissance Hermeticism and astrology was put to practical use, see Anthony Grafton, *Cardano's Cosmos: The Worlds and Works of a Renaissance Astrologer* (Cambridge, Mass.: Harvard University Press, 1999).

18. See Chapter 2, "The New Astronomy and the New Metaphysics," in Alexandre Koyré, *From the Closed World to the Infinite Universe* (Baltimore, Md.: Johns Hopkins University Press, 1957), 28–57.

19. Gershom Scholem, *Major Trends in Jewish Mysticism* (New York: Schocken, 1995 [1941]); Jean-Claude Schmitt, *Le corps, les rites, les rêves, le temps: Essais d'anthropologie medieval* (Paris: Gallimard, 2001); Michael Allen, ed. and trans., *Marsilio Ficino and the Phaedran Charioteer* (Berkeley: University of California Press, 1981); Caroline Walker Bynum, *Wonderful Blood: Theology and Practice in Late Medieval Northern Germany and Beyond* (Philadelphia: University of Pennsylvania Press, 2007).

20. D. P. Walker, *Spiritual and Demonic Magic from Ficino to Campanella* (London: Warburg Institute, 1958), 173–75.

21. See Chapter 10, "Some Renaissance Christian Humanists and 'Superstition,'" and Chapter 11, "Magic, the Fallen World, and Fallen Humanity: Luther on the Devil and Superstitions," in Euan Cameron, *Enchanted Europe: Superstition, Reason, and Religion, 1250–1750* (Oxford: Oxford University Press, 2010), 146–73.

22. Sylvester Prierias, *De strigimagarum daemonumque mirandis* (Rome, 1575).

23. Maggi, *Satan's Rhetoric*; 21–53.

24. Christopher S. Mackay, *The Hammer of Witches: A Complete Translation of the "Malleus Maleficarum"* (Cambridge, U.K.: Cambridge University Press, 2009 [1487]).

25. Johannes Nider, *Formicarius*, Bk. 5 repr. in *Malleus Maleficarum* (1669 ed.), ca. 1435–37.

26. Gustav Freytag, *Pictures of German Life in the Fifteenth, Sixteenth, and Seventeenth Centuries* (London: Chapham and Hall, 1862). Christensen's intertitle here incorrectly cites the title as "A German Life in Pictures."

27. Called a "council" in the title card ("råd" in Swedish).

28. Désiré-Magloire Bourneville and Edmond Teinturier, *Le sabbat des sorciers (Bibliothèque diabolique)* (Paris: Progrès Médical, 1882).

29. Jonathan Crary, *Techniques of the Observer: On Vision and Modernity in the Nineteenth Century* (Cambridge, Mass.: MIT Press, 1991); Alison Griffiths, *Wondrous Difference: Cinema, Anthropology, and Turn-of-the-Century Visual Culture* (New York: Columbia University Press, 2002).

30. Michael Chanan, *The Politics of Documentary* (London: British Film Institute, 2007), 59.

31. Anton Kaes, *Shell Shock Cinema: Weimar Culture and the Wounds of War* (Princeton, N.J.: Princeton University Press, 2009), 31–33.

32. Griffiths, *Wondrous Difference*, 171–84; Chanan, *Politics of Documentary*, 59–74.

33. Brian Winston, *Claiming the Real II: Documentary: Grierson and Beyond* (London: British Film Institute, 2008), 11–13.

34. Ibid.

35. Charlie Kiel, "Steel Engines and Cardboard Rockets: The Status of Fiction and Nonfiction in Early Cinema," in *F Is for Phony: Fake Documentary and Truth's Undoing*, ed. Alexandra Juhasz and Jesse Lerner (Minneapolis: University of Minnesota Press, 2006), 39–49.

36. Mary Ann Doane, *The Emergence of Cinematic Time: Modernity, Contingency, the Archive* (Cambridge, Mass.: Harvard University Press, 2002), 24. See also Tom Gunning, "An Aesthetic of Astonishment: Early Film and the (In)Credulous Spectator," in *Viewing Positions: Ways of Seeing Film*, ed. Linda Williams (New Brunswick, N.J.: Rutgers University Press), 114–33; and Wanda Strauven's edited volume, *The Cinema of Attractions, Reloaded* (Amsterdam: Amsterdam University Press, 2006).

37. Joseph Leo Koerner, *The Reformation of the Image* (Chicago: University of Chicago Press, 2004), 26.

38. For a fuller discussion of Warburg's project see Philippe-Alain Michaud, *Aby Warburg and the Image in Motion* (New York: Zone Books, 2004). Regarding Sherman and Sugimoto, see Michael Fried, *Why Photography Matters as Never Before* (New Haven, Conn.: Yale University Press, 2008).

Marker's singular film is fully analyzed in Alice Harbord, *Chris Marker: "La Jetée"* (London: Afterall Books, 2009).

39. Although largely forgotten today, Esfir Shub was a prominent Soviet filmmaker during the 1920s. Using stock and found footage (including the "home movies" of Tsar Nicholas II, discovered in a cellar in Leningrad), Shub works to reassemble clichéd, amateur images in the expression of both a Marxist historiography and a Bolshevik "creation myth" in *The Fall of the Romanov Dynasty*. Like Christensen, she begins with figurative givens and works, mainly through the montage editing strategies that Soviet cinema of the period is famous for, to empty them out and create new affects and meanings in place of the previous clichés. Michael Chanan summarizes her achievement with this film as "the creation of a film-historical discourse which transcends the simple present tense of the actuality camera which took the original footage." *Politics of Documentary*, 91.

40. Christa Blümlinger notes that Farocki's visual method often draws him to working with both found footage and texts by other authors in order to produce new images and meanings. She describes Farocki's films as "allegorical in that they incorporate fragments and remains. They double or re-read such texts, also making use of second-hand material on the soundtrack, they shoot new images and combine all of these elements afresh, in order to create a 'new analytical capacity of the image' in the sense intended by Deleuze." With the exception of the soundtrack elements, we are ascribing a similar strategy to Christensen in his method of image creation in *Häxan*. Blümlinger, "Slowly Forming a Thought while Working on Images," in *Harun Farocki: Working on the Sightlines*, ed. Thomas Elsaesser (Amsterdam: Amsterdam University Press, 2004), 169.

41. Lorraine Daston and Peter Galison, *Objectivity* (New York: Zone Books, 2007), 37.

42. Hans Vintler, a Tyrolian legal official from a noble family, originally wrote the poem *Buch der Tugend* in 1411. In the poem he laments the widespread popular belief at the time in magic and sorcery. Johann Geiler of Kaiserberg was one of Strasbourg's most popular preachers in the early sixteenth century. In 1509 he delivered a series of sermons on witchcraft and popular magic that were compiled and published under the title *Die Emeis* (*The Ants*) eight years later. See Zika, *Appearance of Witchcraft*, 13, 40–46.

43. On Christensen's part, in his depiction of the Wild Ride, there is perhaps some allusion here to the aerial bombardments of the First World War—the chaos they produced and their force to upend nature. See Kaes, *Shell-Shock Cinema*.

44. Wolfgang Behringer, *Shaman of Oberstdorf: Chonrad Stoeckhlin and the Phantoms of the Night* (Charlottesville: University Press of Virginia, 1998),

26–34. See also Carlo Ginzburg, *The Night Battles: Witchcraft and Agrarian Cults in the Sixteenth and Seventeenth Centuries* (London: Routledge, 1983).

45. Zika, *Appearance of Witchcraft*, 109.

46. Gabriel Audisio, *The Waldensian Dissent: Persecution and Survival, c.1170–c.1570* (Cambridge, U.K.: Cambridge University Press, 1999).

47. Kiel, "Steel Engines and Cardboard Rockets," 46.

48. We have formed this interpretation based on the very strong visual links Christensen highlights in this short series in relation to the visual and historical record of the time. Although a substantially weaker claim in our view, there are also some distant resonances between the unnamed figure in these scenes and the biblical Witch of Endor tale (1 Samuel 28: 3–20). While this female conjuror figures prominently in the *Daemonology* (1597) of King James I of England (King James VI of Scotland) and in William Shakespeare's *Macbeth* (1606), much of the visual art explicitly conjoining the Witch of Endor specifically with witchcraft (rather than with "conjuring" more generally) does not appear to strongly correspond with the scene Christensen has realized in *Häxan*. The clear exception to this is the resonance between the scene and Jacob Cornelisz van Oostsanen's 1526 painting *The Witch of Endor*, which depicts the witch in a magic circle and with a staff, much like the character we witness in *Häxan*. More typically, however, visual works making this link do not so closely resemble Christensen's scene. For example, Daniel Gardiner's painting *The Three Witches from Macbeth* (1775) is a very well-known work of this type, but stylistically Gardiner's depiction is quite far from what the viewer sees in the film. Thus, while Christensen's own images are often quite promiscuous in terms of models and inspiration, by our reading it seems much more likely that the unnamed figure here is referring to Circe rather than the Witch of Endor. For more regarding the biblical figure of the Witch of Endor and the visual culture of witchcraft, see Deanna Petherbridge, "Unholy Trinities and the Weird Sisters of Macbeth," in *Witches and Wicked Bodies* (Edinburgh: National Galleries of Scotland, 2013), 83–98. See also Zika, *Appearance of Witchcraft*, 156–62.

49. Zika, *Appearance of Witchcraft*, 133–41. See also Caroline Walker Bynum, *Metamorphosis and Identity* (New York: Zone Books, 2001).

50. Although Cranach took a great deal of inspiration from Albrecht Dürer's earlier engraving *Melencolia I* (1514), the sense of this earlier work is much different than that found in Cranach's later renditions of the melancholic. As Frances Yates puts it, "Dürer's Melancholy is not in a state of depressed inactivity. She is in an intense visionary trance, a state guaranteed against demonic intervention by angelic guidance. She is not only inspired by Saturn as the powerful star-demon, but also by the angel of Saturn, a spirit with wings like the wings of time." Yates, *Occult Philosophy in the Elizabethan Age*, 66. Considering the trajectory of the developing witch stereotype in the

sixteenth century, it seems that both Dürer's active and Cranach's passive melancholic would selectively come to mind for theologians and artists concerned with the issue, albeit Cranach's version would win out over time. More directly taking up Luther's understanding of melancholy as Satan's "medium for snatching honest souls" than Dürer, Steven Ozment remarks that "Cranach's figures of melancholy are souls in extremis, their only hope of release from such bondage the clarity of the Gospel and the stamina of their faith." In Steven Ozment, *The Serpent and the Lamb: Cranach, Luther, and the Making of the Reformation* (New Haven, Conn.: Yale University Press, 2011), 169.

51. Aby M. Warburg, *Images from the Region of the Pueblo Indians of North America* (Ithaca, N.Y.: Cornell University Press, 1995), 19.

52. Avital Ronell, *The Telephone Book: Technology, Schizophrenia, Electric Speech* (Lincoln: University of Nebraska Press, 1991), 366.

53. Daston and Galison, *Objectivity*, 19.

54. Michaud, *Aby Warburg and the Image in Motion*, 278.

55. Doane, *Emergence of Cinematic Time*, 22.

56. Kaes, *Shell Shock Cinema*, 86.

57. James Clifford, *The Predicament of Culture: Twentieth-Century Ethnography, Literature, and Art* (Cambridge, Mass.: Harvard University Press, 1988), 131.

58. As we noted in the introduction, the ambiguous ground for such "surrealist" documentations of the real was embedded in the very idea of anthropology from the late nineteenth-century forward. Malinowski's own writings display overtones of precisely the qualities the surrealists were trying to bring into view, giving their robust interest in academic ethnographic works a strong logical basis.

59. Clifford, *Predicament of Culture*, 134.

60. For a full discussion of the links between surrealism and ethnography that we are referring to here see Russell, "Surrealist Ethnography." See also Denis Hollier, "The Question of Lay Ethnography [The Entropological Wild Card]," in *Undercover Surrealism: Georges Bataille and* DOCUMENTS, ed. Dawn Ades and Simon Baker (Cambridge, Mass.: MIT Press, 2006), 58–64; and Paul Hammond, ed., *The Shadow and Its Shadow: Surrealist Writings on the Cinema*, 3rd ed. (San Francisco: City Lights Books, 2000).

61. David Bordwell, *The Films of Carl-Theodor Dreyer* (Berkeley: University of California Press, 1981), 28.

62. Kaes, *Shell Shock Cinema*, 85. See also Henri Lefebvre, *The Production of Space* (Oxford, U.K.: Blackwell, 1991).

63. Ulrich Baer, *Spectral Evidence: The Photography of Trauma* (Cambridge, Mass.: MIT Press, 2002), 24.

64. Daston and Galison, *Objectivity*, 39–42.

65. Koerner, *Reformation of the Image*, 141–42. See also Joseph Leo Koerner, "Hieronymus Bosch's World Picture," in *Picturing Science, Producing Art*, ed. Peter Galison and Carolyn Jones (New York: Routledge, 1998), 297–323.

66. Gilles Deleuze, *Francis Bacon: The Logic of Sensation*, trans. Daniel W. Smith (Minneapolis: University of Minnesota Press, 2002), 71–75.

67. Malin Wahlberg, *Documentary Time: Film and Phenomenology* (Minneapolis: University of Minnesota Press, 2008), 3–21.

2. EVIDENCE, SECOND MOVEMENT: TABLEAUX AND FACES

1. Richard Kieckhefer, *Magic in the Middle Ages* (Cambridge, U.K.: Cambridge University Press, 1989), 176–201. See also Hans Peter Broedel, *The "Malleus Maleficarum" and the Construction of Witchcraft: Theology and Popular Belief* (Manchester, U.K.: Manchester University Press, 2003).

2. Modern audiences tend to read Christensen's depiction of the priest in this scene as a kind of crude, anachronistic slapstick. Grotesque figures of this kind have a long history in European popular art and culture, however, dating back to the irreverent, satirical folk humor of the Renaissance carnival. The stark contrast between high and low expressed in one and the same character is not only consistent with the naturalism we detect in *Häxan* but also anticipates the findings of Russian literary theorist Mikhail Bakhtin's classic study on François Rabelais's scandalous satires (particularly *Gargantua and Pantagruel*) and the popular humor evident in the carnival in Renaissance Europe. See Mikhail Bakhtin, *Rabelais and His World*, trans. Hélène Iswolsky (Bloomington: Indiana University Press, 1984 [1968]). In particular, see the chapter "Banquet Imagery in Rabelais," 278–302. See also François Rabelais, *Gargantua and Pantagruel*, trans. M. A. Screech (London: Penguin, 2006 [1532–64]).

3. For an account of Luther's "bad language" see Antónia Szabari, "The Scandal of Religion: Luther and Public Speech in the Reformation," in *Political Theologies: Public Religion in a Post-secular World*, ed. Hent de Vries and Lawrence E. Sullivan (New York: Fordham University Press, 2006), 122–36; see also Szabari, *Less Rightly Said: Scandals and Readers in Sixteenth-Century France* (Stanford, Calif.: Stanford University Press, 2009). A concise analysis of the relation between religious polemics and pornography is offered by Margaret C. Jacob, "The Materialist World of Pornography," in *The Invention of Pornography: Obscenity and the Origins of Modernity, 1500–1800*, ed. Lynn Hunt (New York: Zone Books, 1993), 157–202.

4. Ib Monty, "Benjamin Christensen in Germany: The Critical Reception of His Films in the 1910s and 1920s," in *Nordic Explorations: Film before 1930*, ed. John Fullerton and Jan Olsson (Sydney: John Libbey, 1999), 48.

5. Kieckhefer, *Magic in the Middle Ages*, 81–83.

6. There are a number of excellent studies of the relation between the veneration of saints and relics and the subsequent emergence of anatomy as a foundational principle of Western medical science, including Katharine Park, *Secrets of Women: Gender, Generation, and the Origins of Human Dissection* (New York: Zone Books, 2006); Andrea Carlino, *Books of the Body: Anatomical Ritual and Renaissance Learning* (Chicago: University of Chicago Press, 1999); and Andrew Cunningham, *The Anatomical Renaissance: The Resurrection of the Anatomical Projects of the Ancients* (Aldershot, U.K.: Scolar Press, 1997).

7. Christina Larner, *Witches and Religion: The Politics of Popular Belief* (Oxford, U.K.: Blackwell, 1984), 56, 87.

8. E. E. Evans-Pritchard, *Witchcraft, Oracles and Magic among the Azande*, abridged ed. (Oxford: Oxford University Press, 1976); and Stuart Clark, *Thinking with Demons: The Idea of Witchcraft in Early Modern Europe* (Oxford, U.K.: Oxford University Press, 1997).

9. Edward Bever, "Witchcraft, Female Aggression, and Power in the Early Modern Community," *Journal of Social History* 35, no. 4 (2002): 955–98.

10. Lyndal Roper, *Witch Craze: Terror and Fantasy in Baroque Germany* (New Haven, Conn.: Yale University Press, 2004); Clark, *Thinking with Demons*. See also Sarah Ferber, *Demonic Possession and Exorcism in Early Modern France* (London: Routledge, 2004); and Brian P. Levack, *The Witch-Hunt in Early Modern Europe*, 3rd ed. (London: Longman, 2006).

11. Joseph Leo Koerner, *The Reformation of the Image* (Chicago: University of Chicago Press, 2004), 150.

12. David Bordwell, *The Films of Carl-Theodor Dreyer* (Berkeley: University of California Press, 1981), 29.

13. Bordwell comments on this question in *The Films of Carl-Theodor Dreyer*, and *Making Meaning: Inference and Rhetoric in the Interpretation of Cinema* (Cambridge, Mass.: Harvard University Press, 1989). Others who have weighed in on the issue include Jean-Louis Comolli, "Filmographie commentée," trans. in Mark Nash, *Dreyer* (London: British Film Institute, 1977); Maurice Drouzy, *Carl Th. Dreyer, né Nilsson* (Paris: Editions du Cerf, 1982); and Casper Tybjerg, "Red Satan: Carl Theodor Dreyer and the Bolshevik Threat," in *Nordic Explorations: Film before 1930*, ed. John Fullerton and Jan Olsson (Sydney: John Libbey, 1999), 19–40.

14. Walter Stephens, *Demon Lovers: Witchcraft, Sex, and the Crisis of Belief* (Chicago: University of Chicago Press, 2002), 35.

15. Catherine Russell, "Surrealist Ethnography: *Las Hurdes* and the Documentary Unconscious," in *F Is for Phony: Fake Documentary and Truth's Undoing*, ed. Alexandra Juhasz and Jesse Lerner (Minneapolis: University of Minnesota Press, 2006), 133.

16. Ibid.

17. Walter Benjamin, "One Way Street," in *Selected Writings, Vol. 1: 1913–1926*, ed. Marcus Bullock and Michael W. Jennings (Cambridge, Mass.: Harvard University Press, 1996), 459.

18. Mary Ann Doane, *The Emergence of Cinematic Time: Modernity, Contingency, the Archive* (Cambridge, Mass.: Harvard University Press, 2002), 109.

19. Ibid., 111.

20. For a full account of Marlowe's reactionary, anti-Hermetic version of the Faust legend, see Chapter 11, "The Reaction: Christopher Marlowe on Conjurors, Imperialists, and Jews," in Frances Yates, *The Occult Philosophy in the Elizabethan Age* (London: Routledge, 2001 [1979], 135–47.

21. Wolfgang Behringer, *Shaman of Oberstdorf: Chonrad Stoeckhlin and the Phantoms of the Night* (Charlottesville: University Press of Virginia, 1998), 99–104. See also Carlo Ginzburg, *The Cheese and the Worms: The Cosmos of a Sixteenth-Century Miller* (London: Routledge, 1980).

22. Broedel, *"Malleus Maleficarum" and the Construction of Witchcraft*, 10–14.

23. Behringer, *Shaman of Oberstdorf*, 157–60.

24. Koerner, *Reformation of the Image*, 84.

25. Angela Dalle Vacche, *Cinema and Painting: How Art Is Used in Film* (Austin: University of Texas Press, 1996), 161–96.

26. Thomas Elsaesser, *Weimar Cinema and After: Germany's Historical Imaginary* (London: Routledge, 2000). See also Lotte Eisner, *The Haunted Screen: Expressionism in the German Cinema and the Influence of Max Reinhardt* (Berkeley: University of California Press, 1973).

27. Bordwell, *Films of Carl-Theodor Dreyer*, 54–59.

28. Gilles Deleuze, *Cinema 1: The Movement-Image*, trans. Hugh Tomlinson and Barbara Habberjam (Minneapolis: University of Minnesota Press, 1986), 92.

29. Ibid., 106

30. David Rudkin, *"Vampyr"* (London: British Film Institute, 2005). See also Mark Nash, *Dreyer* (London: British Film Institute, 1977).

31. Deleuze, *Cinema 1*; Bordwell, *Films of Carl-Theodor Dreyer*; Dalle Vacche, *Cinema and Painting*; and Anton Kaes, *Shell Shock Cinema: Weimar Culture and the Wounds of War* (Princeton, N.J.: Princeton University Press, 2009) each discuss the innovative uses of facial close-up shots and tableaux-like images linking *life* to *environment* in many of the best films of the period.

32. Deleuze, *Cinema 1*, 124.

33. Clark, *Thinking with Demons*, 165.

34. Gilles Deleuze, *Francis Bacon: The Logic of Sensation*, trans. Daniel W. Smith (Minneapolis: University of Minnesota Press, 2002), 129.

35. Bordwell, *Films of Carl-Theodor Dreyer*, 66–67; Deleuze, *Cinema 1*, 102–7.

3. THE VIRAL CHARACTER OF THE WITCH

1. Franz Heinemann, *Der Richter und die Rechtspflege in der deutschen Vergangeheit: Mit 159 Abbildungen und Beilagen nach den Originalen aus dem fünfzehnten bis achtzehnten Jahrhundert* (Leipzig: Eugen Diederichs, 1900).

2. Eduard Fuchs, *Illustrierte Sittengeschichte vom Mittelalter bis zur Gegenwart* (Munich: Langen, 1912).

3. See Montague Summers, ed., *The Discovery of Witches: A Study of Master Matthew Hopkins Commonly Called Witch Finder General 1647* (Whitefish, Mont.: Kessinger, 2010).

4. Mary Ann Doane, *The Emergence of Cinematic Time: Modernity, Contingency, the Archive* (Cambridge, Mass.: Harvard University Press, 2002), 125.

5. Brian Winston, *Claiming the Real II: Documentary: Grierson and Beyond* (London: British Film Institute, 2008), 107–12. For studies devoted entirely to Robert Flaherty see Richard Barsam, *The Vision of Robert Flaherty: The Artist as Myth and Filmmaker* (Bloomington: Indiana University Press, 1988); and Paul Rotha, *Robert J. Flaherty: A Biography*, ed. Jay Ruby (Philadelphia: University of Pennsylvania Press, 1983).

6. Charles Zika, *Exorcising Our Demons: Magic, Witchcraft and Visual Culture in Early Modern Europe* (Leiden: Brill, 2003), 214–16.

7. Raymond Klibansky, Erwin Panofsky, and Fritz Saxl, *Saturn and Melancholy: Studies on the History of Natural Philosophy, Religion, and Art* (New York: Basic Books, 1964), 383, specifically on the connection between hordes of witches under Satan's command and the "Saturnine" melancholic.

8. Wolfgang Behringer, *Shaman of Oberstdorf: Chonrad Stoeckhlin and the Phantoms of the Night* (Charlottesville: University Press of Virginia, 1998).

9. Henri Boguet, *An Examen of Witches* (London: Richard Clay and Sons, 1929 [1610]).

10. Michel de Certeau, *The Possession at Loudun*, trans. Michael B. Smith (Chicago: University of Chicago Press, 2000 [1986]), 109–21. See also Lorraine Daston and Peter Galison, *Objectivity* (New York: Zone Books, 2007).

11. This key character is identified as "Anna's sister" in *Häxan*'s official credits. As the dialogue in *Häxan*'s intertitles refers to the character as "the young maiden" at several points, we have adopted this somewhat less impersonal usage to refer to her. As this usage is standing in for a proper name, we have capitalized "Young Maiden" throughout when referring to this character. Popular contemporary sources regarding *Häxan* betray some confusion on the identities of these characters, as sources such as the Internet Movie Database incorrectly refer both to the character played by Astrid Holm (Anna, the Printer's Wife) and to Karen Winther (Anna's sister/the Young Maiden) as simply "Anna." Internet Movie Database, http://www.imdb.com/title/tt0013257/.

12. Jeanne Favret-Saada, *Deadly Words: Witchcraft in the Bocage* (Cambridge, U.K.: Cambridge University Press, 1980); and James Siegel, *Naming the Witch* (Stanford, Calif.: Stanford University Press, 2006) both deal extensively with this procedural logic of witchcraft in ethnographic contexts.

13. Although not played for humor, the sight of Maria's uncouth table etiquette clearly parallels the earlier scene of Father Henrik's gluttony in Chapter 2. Functioning as a visual rhyme linking priest (high) and a soon-to-be accused witch (low), Christensen's citation of the grotesque bodies common in Renaissance and early modern popular culture serves to extend his own form of cinematic naturalism here.

14. Norman Cohn, *Europe's Inner Demons: An Enquiry Inspired by the Great Witch-Hunt* (London: Chatto Heinemann, 1975), 160–63.

15. Lyndal Roper, *Witch Craze: Terror and Fantasy in Baroque Germany* (New Haven, Conn.: Yale University Press, 2004), 136–37.

16. Norman Cohn, *The Pursuit of the Millennium: Revolutionary Millenarians and Mystical Anarchists of the Middle Ages* (London: Pimlico, 2004 [1970]).

17. Margaret Alice Murray, *The Witch-Cult in Western Europe: A Study in Anthropology* (Oxford, U.K.: Clarendon Press, 1921).

18. Cohn, *Europe's Inner Demons*, 160–63.

19. See Dan Edelstein, *The Terror of Natural Right: Republicanism, the Cult of Nature, and the French Revolution* (Chicago: University of Chicago Press, 2010).

20. Max Weber, *The Protestant Ethic and the Spirit of Capitalism*, trans. Talcott Parsons (London: Routledge, 2001 [1930]). See also Alan Macfarlane, *Witchcraft in Tudor and Stuart England: A Regional and Comparative Study* (New York: Harper and Row, 1970); and Keith Thomas, *Religion and the Decline of Magic* (New York: Scribner, 1971).

21. Lyndal Roper directly takes up the Freudian implications of the relations between accuser, accused, and inquisitor in *Witch Craze* and in her earlier study, *Oedipus and the Devil: Witchcraft, Sexuality, and Religion in Early Modern Europe* (London: Routledge, 1994).

22. Walter Stephens, *Demon Lovers: Witchcraft, Sex, and the Crisis of Belief* (Chicago: University of Chicago Press, 2002), 87–124.

23. Stuart Clark, *Thinking with Demons: The Idea of Witchcraft in Early Modern Europe* (Oxford, U.K.: Oxford University Press, 1997), 572–81.

4. DEMONOLOGY

1. Sigmund Freud, *Beyond the Pleasure Principle*, trans. James Strachey (New York: Bantam, 1959 [1920]), and "A Note upon the 'Mystic Writing-Pad,'" in *Collected Papers*, vol. 5, ed. Philip Rieff (New York: Collier, 1963 [1925]). See also Mary Ann Doane, *The Emergence of Cinematic Time:*

Modernity, Contingency, the Archive (Cambridge, Mass.: Harvard University Press, 2002).

2. Lyndal Roper, *Oedipus and the Devil: Witchcraft, Sexuality, and Religion in Early Modern Europe* (London: Routledge, 1994), 206.

3. Hans Peter Broedel, *The "Malleus Maleficarum" and the Construction of Witchcraft: Theology and Popular Belief* (Manchester, U.K.: Manchester University Press, 2003), 32–33.

4. Walter Stephens, *Demon Lovers: Witchcraft, Sex, and the Crisis of Belief* (Chicago: University of Chicago Press, 2002), 180–90.

5. Stuart Clark, *Thinking with Demons: The Idea of Witchcraft in Early Modern Europe* (Oxford, U.K.: Oxford University Press, 1997), 572–81.

6. David Bordwell, *The Films of Carl-Theodor Dreyer* (Berkeley: University of California Press, 1981), 41–53.

7. E. H. Gombrich, *The Story of Art* (New York: Phaidon, 1966), 324. See also Bordwell, *Films of Carl-Theodor Dreyer*, 43.

8. Although there is no overt indication that Christensen intended this, naming his executioner "Erasmus" does lend the character some additional subtext. On one hand, it could simply be a reference to Erasmus of Formia, a venerated saint and martyr who died in 303 (also known as St. Elmo). On the other hand, this namesake may be Erasmus of Rotterdam (1466–1536), most famous today for *Praise of Folly*, a work satirizing various popes and the Church through what Antony Levi has described as "free use of paradox, of ironically exaggerated learning, [and] light-hearted banter." Either way, the naming choice does raise an eyebrow. This is particularly true in light of the fact that the character serves as the literal instrument of the Church's violence against accused witches. While the joke may have passed most audience members by, it can be read as an example both of Christensen's obvious criticism of this violence and his frankly wicked sense of humor. See Anthony Levi, *Renaissance and Reformation: The Intellectual Genesis* (New Haven, Conn.: Yale University Press, 2002), 201. See also Erasmus, *Praise of Folly*, trans. Betty Radice (London: Penguin, 1971 [1511]).

9. Gilles Deleuze, *Cinema 1: The Movement-Image*, trans. Hugh Tomlinson and Barbara Habberjam (Minneapolis: University of Minnesota Press, 1986), 107.

10. Lyndal Roper, *Witch Craze: Terror and Fantasy in Baroque Germany* (New Haven, Conn.: Yale University Press, 2004), 44.

11. Clark, *Thinking with Demons*, 11–30.

12. After Christensen himself, the actor most often singled out from *Häxan*'s ensemble cast is the amateur Maren Pedersen, who plays the withered old Maria. Jack Stevenson sums up her unique presence in the film when he writes, "Years later Christensen would talk about how inspiring it had been to work with Maren Pedersen. She brought 'the real thing' with her into the

studio; the kind of superstitious nature and belief in mysticism that had been commonplace during the Middle Ages. All that the film was about was still alive in her fragile body." Stevenson, *Witchcraft through the Ages: The Story of "Häxan," the World's Strangest Film and the Man Who Made It* (Goldaming, U.K.: FAB Press, 2006), 33.

13. Stephens, *Demon Lovers*, 87–93; Roper, *Witch Craze*, 46.

14. Sigmund Freud, *An Outline of Psychoanalysis* (New York: Norton, 1949 [1940]). See also Carlo Ginzburg, *Myths, Emblems, and Clues* (London: Hutchinson Radius, 1990); and Nick Tosh, "Possession, Exorcism and Psychoanalysis," *Studies in History and Philosophy of Biological and Biomedical Sciences* 33 (2002): 583–96.

15. The specificity of Nordic understandings and expressions of witchcraft is discussed in detail in E. William Monter, "Scandinavian Witchcraft in Anglo-American Perspective," 425–34, and Robert Rowland, " 'Fantasticall and Devilishe Persons': European Witch-Beliefs in Comparative Perspective," 161–90, both appearing in *Early Modern European Witchcraft: Centres and Peripheries*, ed. Bengt Ankarloo and Gustav Henningsen (Oxford, U.K.: Oxford University Press, 1993). Stephen A. Mitchell's *Witchcraft and Magic in the Nordic Middle Ages* (Philadelphia: University of Pennsylvania Press, 2010) is also an excellent source on this topic.

16. St. Thomas Aquinas, *Summa theologiae*, ed. Thomas Gilby et al., 60 vols. (New York: McGraw Hill, 1964), 1.51.2 (9:37). See also Stephens, *Demon Lovers*, 62.

17. Francesco Maria Guazzo, *Compendium maleficarum*, ed. Montague Summers, trans. E. A. Ashwin (New York: Dover, 1988 [1929]), 31. Note that Ashwin translated the first edition of the original text, published in Milan in 1608.

18. Broedel, *"Malleus Maleficarum" and the Construction of Witchcraft*, 40–65.

19. Stephens, *Demon Lovers*, 302–5; Moira Smith, "The Flying Phallus and the Laughing Inquisitor: Penis Theft in the Malleus Maleficarum," *Journal of Folklore Research* 39, no. 1 (2002): 85–117.

20. Armando Maggi, *Satan's Rhetoric: A Study of Renaissance Demonology* (Chicago: University of Chicago Press, 2001), 21–53.

21. Quoted in Arthur Calder-Marshall, *The Innocent Eye: The Life of Robert Flaherty* (New York: W. H. Allen, 1963), 97.

22. Roper, *Witch Craze*, 127–59.

23. Norman Cohn, *Europe's Inner Demons: An Enquiry Inspired by the Great Witch-Hunt* (London: Chatto Heinemann, 1975), 220. See also Joseph Leo Koerner, *The Moment of Self-Portraiture in German Renaissance Art* (Chicago: University of Chicago Press, 1997), 237.

24. Cohn, *Europe's Inner Demons*.

25. Although superimposition of this sort was well established as a special effect in cinema by 1922, it is striking to note the resemblance between Christensen's sequence here and the popular and controversial images produced by Victorian spirit photography during the latter half of the nineteenth century. Jennifer Tucker explains that "Victorian spirit photography was a subgenre in which 'spirits' materialized on photographic plates, often as veiled figures and occasionally as geometric shapes, angels, or blurs" (66). Tucker notes that spirit photography proved a serious problem for proponents of mechanical neutrality in that, on the one hand, the camera was supposed to be a vehicle for purely factual, "subjectless" evidence of the real, and yet on the other seemed to objectively document forces and beings held to be illusions rooted in superstition. These images generated widespread debate over both the existence of such supernatural beings and the trustworthiness of the camera in producing scientific evidence and data, particularly in light of the fact that photography was opening up other invisible worlds (in astronomy and microbiology, for example) that were nevertheless held to be "real." Tucker, *Nature Exposed: Photography as Eyewitness in Victorian Science* (Baltimore, Md.: Johns Hopkins University Press, 2005), 65–125.

26. Cohn, *Europe's Inner Demons*. See also Charles Zika, *The Appearance of Witchcraft: Print and Visual Culture in Sixteenth-Century Europe* (London: Routledge, 2008), 106–9.

27. Cohn, *Europe's Inner Demons*. See also Carlo Ginzburg, *The Night Battles: Witchcraft and Agrarian Cults in the Sixteenth and Seventeenth Centuries* (London: Routledge, 1983), for a more critical reading of Cohn.

28. Zika, *Appearance of Witchcraft*, 99–124.

29. It is clear that Christensen's visualization of the Sabbat corresponds to well-known works from the early seventeenth century that depicted the demonic rites as a carnival of sorts. In particular, these scenes in *Häxan* appear to take some direct inspiration from Jean (Jan) Ziarnko's engraving *Description et figure du sabbat* (1613) and the Michael Herr (or Heer) broadsheet etching (etched by Matthaeus Merian the Elder) *Zauberey; or, Sabbath on the Blocksberg* (1626). Both of these works were derived from Jacques de Gheyn's *Preparation for the Witches' Sabbath* (engraved by Andries Jacobsz Stock, 1610), and Ziarnko's engraving was featured as part of Pierre de Lancre's text *On the Inconstancy of Witches* (*Tableau de l'inconstance des mauvais anges et demons*, 1613).

30. Roper, *Witch Craze*, 111.

31. Zika, *Appearance of Witchcraft*, 133–41.

32. See Part II, Question I, Chapter VIII in the *Malleus Maleficarum*. See Christopher S. Mackay, *The Hammer of Witches: A Complete Translation of the "Malleus Maleficarum"* (Cambridge, U.K.: Cambridge University Press, 2009 [1487]), 332.

33. Stefanos Geroulanos, "A Child Is Being Murdered," in *anthropologies*, ed. Richard Baxstrom and Todd Meyers (Baltimore, Md.: Creative Capitalism, 2008), 17–30.

34. Alphonsus de Spina, *Fortalitium fidei* (Strasbourg, ca. 1471).

35. Clark, *Thinking with Demons*, 81–82.

36. Stephens, *Demon Lovers*, 197–202.

37. Norman Cohn, *The Pursuit of the Millennium: Revolutionary Millenarians and Mystical Anarchists of the Middle Ages* (London: Pimlico, 2004 [1970]), 74–84. See also Stephens, *Demon Lovers*, 145–79.

38. Cohn, *Europe's Inner Demons*, 8–9.

39. 1 Corinthians 11: 23–29.

40. Roper, *Witch Craze*, 81. See also Sigmund Freud, *Totem and Taboo: Some Points of Agreement between the Mental Lives of Savages and Neurotics* (New York: Norton, 1950 [1913]).

41. Cohn, *Pursuit of the Millennium*, and *Europe's Inner Demons*. See also Gabriel Audisio, *The Waldensian Dissent: Persecution and Survival, c.1170–c.1570* (Cambridge, U.K.: Cambridge University Press, 1999); and Malcolm Barber, *The Cathars: Dualist Heretics in Languedoc in the High Middle Ages* (London: Longman, 2000).

42. Charles Zika, *Exorcising Our Demons: Magic, Witchcraft and Visual Culture in Early Modern Europe* (Leiden: Brill, 2003), 445–80.

43. Ibid.

44. See Raymond Klibansky, Erwin Panofsky, and Fritz Saxl, *Saturn and Melancholy: Studies on the History of Natural Philosophy, Religion, and Art* (New York: Basic Books, 1964).

45. Clark, *Thinking with Demons*, 82.

46. Stephens, *Demon Lovers*, 167.

47. *Malleus Maleficarum*, Part III, Q. 15.

48. Roper, *Witch Craze*, 52, 58.

1922

1. Kevin Jackson, in his *Constellation of Genius: 1922, Modernism Year One* (New York: Farrar, Straus and Giroux, 2013), dedicates a few pages to Malinowski, Flaherty, and Christensen, noting that Freud was also writing "A Seventeenth-Century Demonological Neurosis" in 1922 (260).

2. Forsyth Hardy, *Grierson on Documentary* (London: Collins, 1946), 111.

3. Rachael Low, *History of the British Film, Vol. 2: 1906–1914* (London: Routledge, 1997 [1949]), 155.

4. Paul Rotha, *Robert J. Flaherty: A Biography*, ed. Jay Ruby (Philadelphia: University of Pennsylvania Press, 1983), 39.

5. In this era of "armchair anthropology" it was not at all uncommon for the visible body to be regarded as the bedrock object of ethnography, as difference on this register was regarded as obvious and its characteristics could be verified by anyone with access to trustworthy images of the native populations under study. While language, customs, and indeed "culture" itself were important considerations, they were often regarded as secondary "indications" of an objective difference that was mapped on bodies. The experience of cultural Otherness by the missionaries, traders, and explorers made for interesting readings of the primary data, but the truth of the matter required facts and distance for the early anthropological theorists. No single type of evidence seemed better suited for maintaining facticity at a distance than photographs, images "seared with reality," able to yield "facts about which there is no question." See Walter Benjamin, "Little History of Photography," in *Selected Writings, Vol. 1, Part 2: 1931–1934*, ed. Michael Jennings et al. (Cambridge, Mass.: Harvard University Press, 1999), 510, 514. The native body seemed to, finally, resolve the problem of the "bad object" when it came to discovering the truth of human nature, human culture, and human life. This faith in the image, however, assumed that one knew what one was looking for. While contemporary historical narratives of the origin of human sciences such as anthropology assert that what was at stake in these early studies was, then as now, an ability to analytically grasp the nature of human difference while simultaneously retaining the singular universality of "the human" or of "man," this commonsense assertion discounts the multiplicity and incoherence that shaped these disciplines. In fact, while the documentation and analysis of the body emerged as the privileged object in the early days of the human sciences, the broader *scientific objective* of such efforts remained quite inchoate, often resembling a form of comparative anatomy more than anything else. It was only with the publication of Edward Burnett Tylor's *Primitive Culture* in 1871 that an anthropological definition of culture emerged, and even this early attempt to distinguish ethnological work from that of the anatomist was heavily indebted to Darwin's evolutionary framework regarding natural selection and the biological evolution of species. Bearing this in mind, it is still fair to argue for the revolutionary nature of Tylor's move to articulate culture as the object of anthropological science, though this move came at the cost of looping this object right back into the epistemic murk of the invisible and the nonsensical. While Tylor himself worked to preserve the seemingly objective distance afforded by the armchair, his powerful move to give anthropology its own object opened the door for a new vision of anthropology that was premised on precisely the kind of experiential, primary contact that would seem to invalidate the objectivity of evidence obtained in this manner. Taking up the challenge of how to *really* see the Other, ethnographers such as Félix-Louis Regnault, Franz Boas, and

Alfred Cort Haddon worked to formulate a new method that would allow the anthropologist to enter this dangerous world of Otherness without sacrificing the demanded objectivity of any scientific researcher. In doing this, researchers such as Boas and Haddon retained one of the most powerful artifacts available to the armchair anthropologists, the photographic image; now, however, they themselves would carry out the creation of these images, following an evolving methodological protocol that Malinowski would hone into a mythological statement of origin and purpose by the year 1922.

6. George Stocking, *After Tylor: British Social Anthropology, 1888–1951* (Madison: University of Wisconsin Press, 1995,) 270–71. Note that Clifford Geertz (cited earlier in the book) later took up Stocking's "I-Witnessing" phrase in his own discussion of Malinowski's narrative strategies.

7. Michael Young, *Malinowski's Kiriwina: Fieldwork Photography, 1915–1918* (Chicago: University of Chicago Press, 1998).

8. Christopher Pinney, *Photography and Anthropology* (London: Reaktion, 2011), 50–62.

9. Bronislaw Malinowski, *Argonauts of the Western Pacific: An Account of Native Enterprise and Adventure in the Archipelagoes of Melanesian New Guinea* (London: Routledge, 2002 [1922]), 3.

10. Pinney, *Photography and Anthropology*, 60–61.

11. The first edition was published in 1921.

12. Sir James George Frazer, *The Golden Bough: A Study in Magic and Religion* (London: Macmillan, 1976 [1880]).

13. Murray, *The Witch-Cult in Western Europe*, 16.

14. Alan Macfarlane, *Witchcraft in Tudor and Stuart England: A Regional and Comparative Study* (New York: Harper and Row, 1970), 10.

15. Keith Thomas, *Religion and the Decline of Magic* (New York: Scribner, 1971), 516.

16. Norman Cohn, *Europe's Inner Demons: An Enquiry Inspired by the Great Witch-Hunt* (London: Chatto Heinemann, 1975), 107–15.

17. T. M. Luhrmann, *Persuasions of the Witch's Craft: Ritual Magic in Contemporary England* (Cambridge, Mass.: Harvard University Press, 1989), 42–54.

18. "The ideas and theories of magical practice are for magicians both assertions about the real world, and 'lets pretend' fantasies about strange powers, wizards, even dragons. Magicians treat these ideas and theories sometimes as factual assertions, sometimes as fantasy, without necessarily defining to themselves where they stand. It is as if they were playing with belief—and yet they take themselves seriously, act on the results of their divinations, talk about the implications of their ideas. What this really means is that they are not very concerned about the objective 'truth' of their beliefs—a nonchalance at variance with modern ideals of rationality." Ibid., 13.

19. This anecdote is generally credited as something Malinowski said in the presence of Mrs. B. Z. Seligman and she is cited as the source. As quoted in Raymond Firth, "Introduction: Malinowski as Scientist and as Man," in *Man and Culture: An Evaluation of the Work of Bronislaw Malinowski*, ed. Raymond Firth (London: Routledge and Kegan Paul, 1957), 6.

20. For a detailed scholarly analysis of *Heart of Darkness*, see Norman Sherry, *Conrad's Western World* (Cambridge, U.K.: Cambridge University Press, 1980).

21. Joseph Conrad, *Heart of Darkness* (New York: Norton, 1988 [1899]), 49–50.

22. For an account of Rivers's life, see Richard Slobodin, *W.H.R. Rivers: Pioneer, Anthropologist, Psychiatrist of "The Ghost Road."* Gloucestershire, U.K.: Sutton, 1997 [1978].

23. The findings of these ethnographic studies were published in several forms during the second decade of the twentieth century. The two major works are W.H.R. Rivers, *Kinship and Social Organization* (London: Constable, 1914), and *The History of Melanesian Society: Percy Sladen Trust Expedition to Melanesia*, 2 vols. (Cambridge, U.K.: Cambridge University Press, 1914). The clear innovation in ethnographic method that these works represent was presaged to a strong degree by an early work focused on South India, *The Todas* (London: Macmillan, 1906).

24. H. Rider Haggard, *King Solomon's Mines* (London: Cassell and Company, 1885).

25. W.H.R. Rivers, *Medicine, Magic and Religion* (London: Routledge, 2001 [1924]), 4, 7, 24–26, 48.

26. W.H.R. Rivers, *Instinct and the Unconscious* (Cambridge, U.K.: Cambridge University Press, 1920). See also *Conflict and Dream* (London: Kegan Paul, 1923).

27. Sigmund Freud, *Totem and Taboo: Some Points of Agreement between the Mental Lives of Savages and Neurotics* (New York: Norton, 1950 [1913]), 174–81.

28. W.H.R. Rivers, "The Repression of War Experience," *Proceedings of the Royal Society of Medicine* (Psychiatry Section) 11 (1918): 1–20.

29. Rivers, *Medicine, Magic and Religion*, 4.

30. Anton Kaes, *Shell Shock Cinema: Weimar Culture and the Wounds of War* (Princeton, N.J.: Princeton University Press, 2009), 97.

5. SEX, TOUCH, AND MATERIALITY

1. Hans Peter Broedel, *The "Malleus Maleficarum" and the Construction of Witchcraft: Theology and Popular Belief* (Manchester, U.K.: Manchester University Press, 2003), 10–14.

2. Alasdair MacIntyre, *After Virtue* (Notre Dame, Ind.: University of Notre Dame Press, 1981). See also Talal Asad, *Genealogies of Religion* (Baltimore, Md.: Johns Hopkins University Press, 1993).

3. Johann Weyer, *De praestigiis daemonium*, in *Witches, Devils, and Doctors in the Renaissance*, edited by George Mora. Binghamton, N.Y.: Medieval and Renaissance Texts and Studies, 1991 [1563].

4. Stuart Clark, *Thinking with Demons: The Idea of Witchcraft in Early Modern Europe* (Oxford, U.K.: Oxford University Press, 1997), 205–13.

5. Talal Asad, *Formations of the Secular: Christianity, Islam, Modernity* (Stanford, Calif.: Stanford University Press, 2003), 67–99.

6. Judith Perkins, *The Suffering Self: Pain and Narrative in the Early Christian Era* (New York: Routledge, 1995), 117.

7. Sigmund Freud, "Obsessive Actions and Religious Practices," in *The Standard Edition of the Complete Psychological Works of Sigmund Freud*, ed. James Strachey with Anna Freud (London: Hogarth Press, 1961 [1907]), 115–27; "A Child Is Being Beaten," in *Sexuality and the Psychology of Love* (New York: Bantam, 1968 [1919]); and *The Future of an Illusion*, trans. James Strachey (New York: Norton, 1975 [1927]).

8. Niklaus Largier, *In Praise of the Whip: A Cultural History of Arousal*, trans. Graham Harman (New York: Zone Books, 2007), 221.

9. Bill Nichols, Chapter 7 (with Christian Hansen and Catherine Needham), "Pornography, Ethnography, and Discourses of Power," in *Representing Reality: Issues and Concepts in Documentary* (Bloomington: Indiana University Press, 1991), 201–28.

10. Ib Monty, "Benjamin Christensen in Germany: The Critical Reception of His Films in the 1910s and 1920s," in *Nordic Explorations: Film before 1930*, ed. John Fullerton and Jan Olsson (Sydney: John Libbey, 1999), 41–55.

11. Sigmund Freud, *On Sexuality: Three Essays on the Theory of Sexuality and Other Works*, trans. James Strachey (London: Penguin, 1985 [1905]). See also "Part III: General Theory of the Neuroses, XXVI: The Libido Theory and Narcissism," in *A General Introduction to Psychoanalysis* (New York: Horace Liveright, 1920), 209–402.

12. Richard von Krafft-Ebing, *Psychopathia Sexualis: Eine Klinisch-Forensische Studie* (Stuttgart: Enke, 1886).

13. Marquis de Sade, *The 120 Days of Sodom, and Other Writings*, trans. Austryn Wainhouse and Richard Seaver (New York: Grove Press, 1966). See also Timo Airaksinen, *The Philosophy of the Marquis de Sade* (London: Routledge, 1995).

14. Gilles Deleuze, "Coldness and Cruelty," in *Masochism* (New York: Zone Books, 1991 [1967]), 17.

15. David Bordwell, *The Films of Carl-Theodor Dreyer* (Berkeley: University of California Press, 1981), 54–65.

16. Manuel do Valle De Moura, *De incantantionibus seu ensalmis* (Lisbon, 1620); Armando Maggi, *Satan's Rhetoric: A Study of Renaissance Demonology* (Chicago: University of Chicago Press, 2001).

17. Maggi, *Satan's Rhetoric*, 54–95.

18. Mary Daly, *Gyn/ecology: The Metaethics of Radical Feminism* (Boston: Beacon, 1978).

19. Starhawk, *The Spiral Dance: A Rebirth of the Ancient Religion of the Great Goddess* (San Francisco: Harper and Row, 1979).

20. Jim Sharpe, "Women, Witchcraft, and the Legal Process," in *Women, Crime, and the Courts in Early Modern England*, ed. Jennifer Kermode and Garthine Walker (Chapel Hill: University of North Carolina Press, 1994), 113–30.

21. Diane Purkiss, *The Witch in History: Early Modern and Twentieth-Century Representations* (London: Routledge, 1996), 8.

22. Clark, *Thinking with Demons*, 112–18. See also Broedel, *"Malleus Maleficarum" and the Construction of Witchcraft*, 167–88.

23. Bordwell, *Films of Carl-Theodor Dreyer*, 139, 194.

24. Katharine M. Rogers, *The Troublesome Helpmate: A History of Misogyny in Literature* (Seattle: University of Washington Press, 1966).

25. Caroline Walker Bynum, *Christian Materiality: An Essay on Religion in Late Medieval Europe* (New York: Zone Books, 2011).

26. Walter Stephens, *Demon Lovers: Witchcraft, Sex, and the Crisis of Belief* (Chicago: University of Chicago Press, 2002), 61–63.

27. Luther provides a detailed explanation of "images of the heart" in his 1525 treatise *Against the Heavenly Prophets in the Matter of Images and Sacraments* (*Widder die hymelischen propheten, von den bildern und Sacrament*), published in two parts in Wittenberg. Koerner takes up the notion in his extended analysis of Lucas Cranach the Elder's *Wittenberg Altarpiece* (1547) and the nature of the images for early Protestants; Koerner's text is our source for Luther's translated phrase. Koerner, *Reformation of the Image*, 160.

28. Bynum, *Christian Materiality*, 31–60.

29. Allan Casebier, *Film and Phenomenology: Towards a Realist Theory of Cinematic Representation* (Cambridge, U.K.: Cambridge University Press, 1991). See also David MacDougall, *The Corporeal Image: Film, Ethnography, and the Senses* (Princeton, N.J.: Princeton University Press, 2006); and Malin Wahlberg, *Documentary Time: Film and Phenomenology* (Minneapolis: University of Minnesota Press, 2008).

30. Walter Benjamin, *The Work of Art in the Age of Its Technological Reproducibility, and Other Writings on Media*, ed. Michael W. Jennings, Brigid Doherty, and Thomas Y. Levin (Cambridge, Mass.: Harvard University Press, 2008).

31. Deleuze, "Coldness and Cruelty."

32. As we mentioned in the introduction, the double bind of, on one hand, needing to make a case for witchcraft (through a logic of expertise more than anything) in order to, on the other, "un-witch" the witch (as it were), is precisely the dilemma Favret-Saada describes in her ethnographic work in contemporary France. See also Jeanne Favret-Saada, *The Anti-Witch*, trans. Matthew Carey (Chicago: HAU Books, 2015 [2009]).

33. Richard Kieckhefer, *Magic in the Middle Ages* (Cambridge, U.K.: Cambridge University Press, 1989), 182–85.

34. This grossly inflated figure for deaths due to the witch hunts in Europe was invented in Étienne-Léon de Lamothe-Langon's *Histoire de l'inquisition en France, depuis son établissement au XIIIe siècle, à la suite de la croisade contre les Albigeois, jusqu'en 1772, époque définitive de sa suppression* (Paris: J.-G. Dentu, 1829). The inaccurate number was further popularized by Joseph Hansen in *Zauberwahn, Inquisition und Hexenprozess im Mittelalter und die Entstehung der grossen Hexenverfolgung* (Munich: R. Oldenbourg, 1900). Richard Kieckhefer in *European Witch Trials: Their Foundations in Popular and Learned Culture, 1300–1500* (London: Routledge, 1976), and Norman Cohn in *Europe's Inner Demons: An Enquiry Inspired by the Great Witch-Hunt* (London: Chatto Heinemann, 1975) definitively exposed the original forgery. The entire affair is well summarized by Malcolm Gaskill in *Witchcraft: A Very Short Introduction* (Oxford, U.K.: Oxford University Press, 2010).

35. Anton Kaes, *Shell Shock Cinema: Weimar Culture and the Wounds of War* (Princeton, N.J.: Princeton University Press, 2009), 82.

36. Bordwell, *Films of Carl-Theodor Dreyer*, 124–25.

6. POSSESSION AND ECSTASY

1. The focus on these instruments is so strong that Kevin Jackson, even in his brief account of the film, notes, despite highlighting the Church's hypocrisy, ignorance, and cruelty, Christensen nevertheless "dwells lovingly on instruments of torture." Kevin Jackson, *Constellation of Genius: 1922, Modernism Year One* (New York: Farrar, Straus and Giroux, 2013), 260.

2. Lucius Apuleius, *Ain schön lieblich auch kurtzweylig gedichte Lucij Apuleij von ainem gulden Esel . . . Mit schönen figuren zůgericht, grundtlich verdeutscht durch Herren Johan Sieder, etc.* Ff. 71. A (Weissenhorn: Augustae Vindelicorum, 1538).

3. Richard Kieckhefer, *Magic in the Middle Ages* (Cambridge, U.K.: Cambridge University Press, 1989), 176–201.

4. Jack Stevenson, *Witchcraft through the Ages: The Story of "Häxan," the World's Strangest Film and the Man Who Made It* (Goldaming, U.K.: FAB Press, 2006), 66.

5. Désiré-Magloire Bourneville, Paul Régnard, Jean-Martin Charcot, and Édouard Delessert, *Iconographie photographique de la Salpêtrière: Service de M. Charcot*, 3 vols. (Paris: Progrès Médical, 1877–80).

6. Ibid.

7. Désiré-Magloire Bourneville, *Science et miracle: Louise Lateau ou la stigmatisée belge (Bibliothèque diabolique)* (Paris: Progrès Médical, 1875); and Désiré-Magloire Bourneville and Edmond Teinturier, *Le sabbat des sorciers (Bibliothèque diabolique)* (Paris: Progrès Médical, 1882).

8. Michel de Certeau *The Possession at Loudun*, trans. Michael B. Smith (Chicago: University of Chicago Press, 2000 [1986]); and Michel Foucault, "Déviations religieuses et savoir médical," in *Hérésies et societies dans l'Europe préindustrielle, 11e–18e siècles*, ed. Jacques LeGoff (Paris: Mouton, 1968), 19–25.

9. Joseph Leo Koerner, *The Reformation of the Image* (Chicago: University of Chicago Press, 2004), 105.

10. Lorraine Daston and Katharine Park, *Wonders and the Order of Nature, 1150–1750* (New York: Zone Books, 1998).

11. De Certeau, *Possession at Loudun*, 35–51.

7. HYSTERIAS

1. Otto Binswanger, *Die Hysterie* (Wien: Hölder, 1904).

2. Mark S. Micale, "Hysteria and Its Historiography: A Review of Past and Present Writings (I and II)," *History of Science* 27, no. 77 (1989): 224–61, 319–51, and *Hysterical Men: The Hidden History of Male Nervous Illness* (Cambridge, Mass.: Harvard University Press, 2008). See also Doris Kaufmann, "Science as Cultural Practice: Psychiatry in the First World War and Weimar Germany," *Journal of Contemporary History* 34, no. 1 (1999): 125–44.

3. Otto Binswanger, "Die Kriegshysterie," in *Handbuch der Ärztlichen Erfahrungen im Weltkriege, 1914–18*, ed. O. Schjerning, vol. 4: *Geistes- und Nevenkrankheiten*, ed. K. Bonhoeffer (Leipzig: J. A. Barth 1922), 45–67. See also Paul Lerner, *Hysterical Men: War, Psychiatry, and the Politics of Trauma in Germany, 1890–1930* (Ithaca, N.Y.: Cornell University Press, 2009); and Paul Fussell, *The Great War and Modern Memory* (New York: Oxford University Press, 2000).

4. Walter Benjamin, "The Storyteller: Observations on the Work of Nikolai Leskov," in *Selected Writings, Vol. 3: 1935–1938*, ed. Howard Eiland and Michael W. Jennings (Cambridge, Mass.: Harvard University Press, 2006), 143–66.

5. Catherine Malabou, *The New Wounded: From Neurosis to Brain Damage* (New York: Fordham University Press, 2012).

6. Margaret Alice Murray, *The Witch-Cult in Western Europe: A Study in Anthropology* (Oxford, U.K.: Clarendon Press, 1921).

7. Christopher S. Mackay, *The Hammer of Witches: A Complete Translation of the "Malleus Maleficarum"* (Cambridge, U.K.: Cambridge University Press, 2009 [1487]), 261. For a different translation of the same passage see Hans Peter Broedel, *The "Malleus Maleficarum" and the Construction of Witchcraft: Theology and Popular Belief* (Manchester, U.K.: Manchester University Press, 2003), 20–21.

8. Sigmund Freud, "Mourning and Melancholy," in *The Standard Edition of the Complete Psychological Works of Sigmund Freud*, ed. James Strachey with Anna Freud (London: Hogarth Press, 1961 [1917]), 14:243–58.

9. Armando Maggi, *Satan's Rhetoric: A Study of Renaissance Demonology* (Chicago: University of Chicago Press, 2001), 137–79. See also Michel de Certeau *The Possession at Loudun*, trans. Michael B. Smith (Chicago: University of Chicago Press, 2000).

10. Georges Didi-Huberman, *Invention of Hysteria: Charcot and the Photographic Iconography of the Salpêtrière*, trans. Alisa Hartz (Cambridge, Mass.: MIT Press, 2003), 277.

11. Ibid.

12. James Siegel, *Naming the Witch* (Stanford, Calif.: Stanford University Press, 2006), 29–52.

POSTSCRIPT: IT IS VERY HARD TO BELIEVE . . .

1. Timothy S. Murphy, *Wising Up the Marks: The Amodern William Burroughs* (Berkeley: University of California Press, 1997), 202, 206.

2. Bryony Dixon, *100 Silent Films* (London: British Film Institute, 2011), 239–41.

3. Alfred Métraux, *Voodoo in Haiti* (New York: Schocken, 1972 [1958]).

4. Maya Deren, *Divine Horsemen: The Living Gods of Haiti* (New York: Thames and Hudson, 1953).

5. Some representative examples would include Walter B. Cannon, "'Voodoo' Death," *American Anthropologist* 44, no. 2 (1942): 169–81; Marlene Dobkin de Rios, "The Relationship between Witchcraft Beliefs and Psychosomatic Illness," in *Anthropology and Mental Health: Setting a New Course*, ed. Joseph Westermeyer (The Hague: Mouton, 1976), 11–18; Janice Boddy, *Wombs and Alien Spirits: Women, Men, and the Zār Cult in Northern Sudan* (Madison: University of Wisconsin Press, 1989).

6. Michel Foucault, *The Order of Things: An Archaeology of the Human Sciences* (New York: Vintage, 1970 [1966]), 320.

7. The phrasing we are using here is actually a paraphrasing of Louis Couperus's description of the character Van Oudijck in his 1900 novel *The Hidden Force*. The colonial official Van Oudijck represents the sober, objective rationality of science and empire in the book. Couperus, drawing on his own

experience in the Dutch East Indies (now Indonesia), posits that this form of reasoning is insufficient in understanding Javanese life or living in a "Javanese" world. See Couperus, *The Hidden Force* (Amherst: University of Massachusetts Press, 1985 [1900]), 129–30.

8. See Birgit Meyer and Dick Houtman, "Introduction: Material Religion—How Things Matter," in *Things: Religion and the Question of Materiality*, ed. Dick Houtman and Birgit Meyers (New York: Fordham University Press, 2012), 1–25; Pamela Reynolds, *Traditional Healers and Childhood in Zimbabwe* (Athens: Ohio University Press, 1996).

9. Gilles Deleuze, *Cinema 1: The Movement-Image*, trans. by Hugh Tomlinson and Barbara Habberjam (Minneapolis: University of Minnesota Press, 1986), 125.

10. Ibid.

11. Gilles Deleuze, *Cinema 2: The Time-Image* (Minneapolis: University of Minnesota Press, 1989). See also John Rajchman, *The Deleuze Connections* (Cambridge, Mass.: MIT Press, 2000); and Paola Marrati, *Gilles Deleuze: Cinema and Philosophy* (Baltimore, Md.: Johns Hopkins University Press, 2008).

12. Gilles Deleuze, "The Brain Is the Screen," in *Two Regimes of Madness: Texts and Interviews, 1975–1995* (New York: Semiotext[e], 2006), 282–91. See also Giuliana Bruno, "Pleats of Matter, Folds of the Soul," in *Afterimages of Gilles Deleuze's Film Philosophy*, ed. D. N. Rodowick (Minneapolis: University of Minnesota Press, 2010), 213–34.

13. Deleuze, *Cinema 2*; T. M. Luhrmann, "A Hyperreal God and Modern Belief: Toward an Anthropological Theory of Mind," *Current Anthropology* 53, no. 4 (2012): 371–95.

Baer, Casimir Hermann. *Deutsche Wohn- & Festräume aus sechs Jahrhunderten (Bauformen-Bibliothek)*. Stuttgart: J. Hoffmann, 1912.

Bénet, Armand. *Proces verbal fait pour délivrer une fille possédée par le malin espirit, a Louviers (Bibliothèque diabolique)*. Paris: Progrès Médical, 1883 [1591].

Berjon, Antoine. *La grande hysterie chez l'homme*. Paris: J.-B. Bailliere et Fils, 1886.

Bourneville, Désiré-Magloire. *La possession de Jeanne Fery; religieuse professe du couvent des Soeurs Noires de la Ville de Mons*. Paris: Progrès Médical, 1886 [1584].

Bourneville, Désiré-Magloire, Paul Régnard, Jean-Martin Charcot, and Édouard Delessert. *Iconographie photographique de la Salpêtrière: Service de M. Charcot*. 3 vols. Paris: Progrès Médical, 1877–80.

Bourneville, Désiré-Magloire, and Edmond Teinturier. *Le sabbat des sorciers (Bibliothèque diabolique)*. Paris: Progrès Médical, 1882.

Brandes, Georg. *William Shakespeare*. Copenhagen: Gyldendalske Boghandels Forlag, 1895.

Bullot, Maximilien. *Histoire des orders monastiques, religieux et militaries et des congregations séculierés de l'un & l'autre sexe, qui ont eté eéablies jusqu'a present*. 8 vols. Paris: Jean Baptiste Coignard, 1721.

Burgl, Georg. *Die Hysterie und die strafrechtliche Verantwortlichkeit der Hysterischen: Ein praktisches Handbuch fur Aerzte und Juristen. Mit zwanzig ausgewahlten Fallen krimineller Hysterie mit Aktenauszug und gerichtlichen Gutachten*. Stuttgart: F. Enke, 1912.

Campbell, John Gregorson. *Witchcraft and Second Sight in the Highlands and Islands of Scotland: Tales and Traditions Collected Entirely from Oral Sources*. Glasgow: James MacLehose and Sons, 1902.

Charcot, J.-M., and Paul Richer. *Les démoniaques dans l'art*. Paris: Delahaye et Lecrosnier, 1887.

Charcot, J.-M., Paul Richer, Georges Gilles de la Tourette, and Albert Londe, eds. *Nouvelle iconographie de la Salpêtrière*. Paris: Lecrosnier et Babé, 1888–1918.

Daugaard, Jacob Brøgger. *Om de danske Klostre I Middelalderen: Et Priisskrift*. Copenhagen: Andreas Seidelin, 1830.

Diederichs, Eugen, ed. *Deutsches Leben der Vergangenheit in Bildern: Ein Atlas mit 1760 Nachbilddungen alter Kupfer—und Holzschnitte aus dem 15.–18. Jahrundert*. 2 vols. Jena: Eugen Diederichs, 1908.

Freud, Sigmund. *Vorlesungen zur Einfuhrung in die Psychoanaluse*. Leipzig: Heller, 1917.

Fuchs, Eduard. *Illustrierte Sittengeschichte*. 6 vols. Munich: Albert Langen, 1909.

Gadelius, Bror. *Tro och ofvertro i gangna tider*. 2 vols. Stockholm: Hugo Gebers Forlag, 1912–13.

Gilles de la Tourette, Georges. *Hypnotismen og de med den beslægtede tilstande i retslægevidenskabelig belysning*. Copenhagen: Emil Bergmanns, 1888.

Graf, Arturo. *Geschichte des teufelsglaubens: Einzig rechtmassige ausgabe*. Jena: Costenoble, 1893.

Grupp, Georg. *Kulturgeschichte des Mitteralters: 2:te, vollst. neue Bearb*. Paderborn: F. Schöningh, 1907–14.

Hansen, Joseph. *Zauberwahn, Inquisition und Hexenprozess im Mittelalter und die Entstehung der grossen Hexenverfolgung (Historische Bibliothek)*. Leipzig: R. Oldenbourg, 1900.

Heegaard, Poul. *Populaer Astronomi*. Copenhagen: Det Nordiske Forlag, 1902.

Heimbucher, Max. *Die Orden und Kongregationen der Katholischen Kirche*. 3 vols. Paderborn: F. Schöningh, 1907–8.

Heinemann, Franz. *Der Richter und die Rechtspflege in der deutschen Vergangenheit*. 12 vols. Edited by Georg Steinhausen. Leipzig: Eugen Diederichs, 1900.

Hirth, Georg. *Kulturgeschichtliches Bilderbuch aus drei Jahrhunderten*. 6 vols. Leipzig: Hirth, 1896.

Høffding, Harald. *Oplevelse og Tydning: Religionsfilosofiske Studier*. Copenhagen: Nyt Nordisk Forlag, 1918.

Jørgensen, Henrik Gothard Antonius. *Fra bispernes og munkenes tid: Billeder fra den Danske kirkes historie i den senere middelalder*. Copenhagen: I Kommission Hos G.E.C. Gad, 1917.

Jung, Carol Gustav. *Psychology of the Unconscious*. Translated by B. M. Hinkle. New York: Moffat Yard and Co., 1916.

Kaarsberg, Hans S. *Om Satanismen: Djaelebesaettelse og Hexenvaesen set fra laegevidenskabeligt Standpunkt*. Copenhagen: Gyldendal, 1896.

Knudsen, Johannes. *Lucidarius: En Folkebog fra Middelalderen Folkelaesning, Nr. 88*. Copenhagen: I Kommission Hos Gad, 1909.

Lacroix, Paul. *Les arts au moyen âge et à l'époque de la renaissance*. Paris: Firmin-Didot, 1871.

————. *Moeurs, usages et costumes au moyen âge et à l'époque de la renaissance.* Paris: Firmin-Didot, 1872.

————. *Sciences and lettres au moyen âge et à l'époque de la renaissance.* Paris: Firmin-Didot, 1877.

————. *Vie militaire et religieuse au Moyen Age et à l'époque de la Renaissance.* Paris: Firmin-Didot, 1873.

Ladame, Paul Louis. *Procès criminel de la dernière sorcière brulée à Genève le 6 avril 1652, pub. d'après des documents inédits et originaux conservés aux Archives de Genève (No. 3465).* Paris: A. Delahaye et Lecrosnier, 1888.

Lecky, William Edward Hartpole. *History of the Rise and Influence of the Spirit of Rationalism in Europe.* 2 vols. London: Longman, Green, Longman, Roberts and Green, 1865.

Lehmann, Alfred. *Hypnosen og de dermed beslaegtede normale Tilstande.* Copenhagen: Philipsen, 1890.

————. *Overtro og trolddom fra de aeldste tider til vore dage.* Copenhagen: J. Frimodts Forlag, 1893–96.

Lenormant, François. *Die geheimwissenschaften Asiens: Die Magie und Wahrsagekunst der Chaldier.* Jena: H. Costenoble, 1878.

Liebe, Georg. *Das Judentum in der deutschen Vergangenheit.* Leipzig: E. Diederichs, 1903.

Linderholm, Emanuel. *De stora häxprocesserna i Sverige: Bidrag till svensk kultur- och kyrkohistoria.* Uppsala: J. A. Lindblads Förlag, 1918.

Maeterlinck, L. *Le genre satirique dans la peinture flamande.* Ghent: Librairie Néerlandaise, 1903.

Mannhart, W. *Zauberglaube und geheimwissen im Spiegel der Jahrhunderte.* Berlin: Barsdorf, 1909.

Maspero, G. *Histoire ancienne des peuples de l'Orient.* Paris: Hachette, 1886.

Matthiessen, Hugo. *Bøddel og galgefugl: Et kulturhistorisk forsøg.* Copenhagen: Gyldendal, 1910.

————. *Natten; studier i gammelt byliv.* Copenhagen: Gyldendal, 1914.

Michelet, Jules. *La sorcière.* Paris: J. Chevrel, 1911.

Myer, Frederic William Henry. *Human Personality and Its Survival of Bodily Death.* London: Longmans Green and Co., 1903.

Nyrop, Kristoffer. *Kysset og dets historie.* Copenhagen: Bojesen, 1897.

Paracelsus. *Oeuvres complètes de Philippe Aureolus Theophraste Bombast de Hohenheim, dit Paracelse.* Translated by Émile Grillot de Givry. Paris: Bibliothèque Chacornac, 1913–14.

Parmentier, A. *Album historique, publié sous la direction de M. Ernest Lavisse.* Paris: A. Colin, 1897–1907.

Perty, Maximillian. *Die mystischen Erscheinungen der menschlichen Natur. Dargestellt und gedeutet von Maximilian Perty.* Leipzig: C. F. Winter, 1872.

Piton, Camille. *Le costume civil en France du XIIIe au XIXe siècle; ouvrage orné de 700 illustrations directes par la photographie, d'apres les documents du temps (statues, peintures murales, tapisseries, vitraux, etc.).* Paris: Ernest Flammarion, 1913.

Placzek, Siegfried. *Das Geschechtsleben der Hysterischen: Eine medizinische, soziologische und forensische Studie.* Bonn: A. Marcus and E. Weber Verlag, 1919.

Platter, Felix. *Felix Platters Ungdoms-erindringer, skildringer fra Basel og Montpellier i reformationstiden.* Copenhagen: Vilhelm Tryde, 1915.

Putman, Allen. *Witchcraft of New England Explained by Modern Spiritualism.* Boston: Colby and Rich, 1881.

Régnard, Paul. *Les maladies épidémiques de l'esprit; sorcellerie, magnétisme, morphinisme, délire des grandeurs.* Paris: E. Plon Nourrit et Cie, 1887.

Riezler, Sigmund. *Geschichte der Hexenprocesse in Bayern: Im Lichte der allgemeinen Entwickelung dargestellt.* Stuttgart, 1896.

Rouby, Hippolyte. *L'hystérie de Sainte Thérèse.* Paris: F. Alcan, 1902.

Rydberg, Viktor. *Middelalderens Magi.* Copenhagen: Schou, 1884.

Schück, Henrik. *Vår förste författare: En själshistoria från medeltiden.* Stockholm: Hugo Gebers Förlag, 1916.

Seligmann, Siegfried. *Der böse blick und verwandtes: Ein beitrag zur geschichte des aberglaubens aller zeiten und völker.* Berlin: H. Barsdorf, 1910.

Sprenger, Jakob, Inquisitor, Heinrich Institoris (J. W. R. Schmidt). *Der Hexenhammer: Malleus Maleficarum.* Berlin: Barsdorf, 1906.

Stenström, Matts A. *Dansen: Dess utveckling från urtiden till danspalatsens tidevarv.* Stockholm: Söderström and Co., 1918.

Storck, Karl. *Der Tanz.* Bielefeld: Velhagen and Klasing, 1903.

Taylor, John M. *The Witchcraft Delusion in Colonial Connecticut, 1647–1697.* New York: Grafton, 1908.

Turberville, Arthur Stanley. *Medieval Heresy and the Inquisition.* London: C. Lockwood and Son, 1920.

Veth, Jan, and Samuel Muller. *Albrecht Dürers niederländische reise.* Berlin: Grote, 1918.

Weyer, Johann [sometimes rendered Jan or Johan Wier]. *De Praestigiis Daemonum, & incantationibus ac veneficiis Libri sex, postrema editione sexta aucti & recogniti.* Basel: Oporinus, 1583. [English translation of *De Praestigiis Daemonum*: Witches, Devils, and Doctors in the Renaissance; in French: *Histoires, disputes et discours des illusions et impostures des diables.*]

Worringer, Wilhelm. *Die Altdeutsche Buchillustration.* Leipzig: R. Piper and Co., 1912.

Antichrist. Lars von Trier. Denmark/Germany/France/Sweden/Italy/Poland, 2009.

The Artist's Dilemma. Edwin S. Porter. U.S., 1901.

As You See [Wie man sieht]. Harun Farocki. West Germany, 1986.

Attack on a China Mission Station. James Williamson. U.K., 1900.

The Battle of the Somme. Geoffrey Malins and John McDowell. U.K., 1916.

Blind Justice [Hævnens nat]. Benjamin Christensen. Denmark, 1916.

The Cabinet of Dr. Caligari [Das Cabinet des Dr. Caligari]. Robert Wiene. Germany, 1920.

The Child [Barnet]. Benjamin Christensen. Denmark, 1940.

Children of Divorce [Skilsmissens Børn]. Benjamin Christensen. Denmark, 1939.

Come Home with Me [Gaa med mig hjem]. Benjamin Christensen. Denmark, 1940.

Cops. Buster Keaton and Edward F. Cline. U.S., 1922.

Day of Wrath [Vredens dag]. Carl Theodor Dreyer. Denmark, 1943.

Dead Birds. Robert Gardner. U.S., 1963.

The Decameron. Pier Paolo Pasolini. Italy, 1971.

Demolition of a Wall [Démolition d'un mur]. Louis Lumière. France, 1896.

Destiny [Der müde Tod]. Fritz Lang. Germany, 1921.

The Devils. Ken Russell. U.K., 1971.

The Devil's Circus. Benjamin Christensen. U.S., 1926.

Divine Horsemen: The Living Gods of Haiti. Maya Deren. U.S., 1977.

Dr. Mabuse, der Spieler [Dr. Mabuse the Gambler]. Fritz Lang. Germany, 1922.

Dracula. Tod Browning. U.S., 1932.

The Fall of the Romanov Dynasty [Padenie dinastii Romanovykh]. Esfir Shub. USSR, 1927.

Faust. F. W. Murnau. Germany, 1926.

Foolish Wives. Erich von Stroheim. U.S., 1922.

Forest of Bliss. Robert Gardner. U.S., 1985.

Frankenstein. J. Searle Dawley. U.S., 1910.

Frankenstein. James Whale. U.S., 1932.

The Haunted House. Benjamin Christensen. U.S., 1928.

The Hawk's Nest. Benjamin Christensen. U.S., 1928.

Häxan. Benjamin Christensen. Sweden/Denmark, 1922.

Histoire(s) du cinema. Jean-Luc Godard. France, 1988–98.

House of Horror. Benjamin Christensen. U.S., 1929.

Hunting Big Game in Africa. Francis Boggs. U.S., 1907.

An Image [*Ein Bild*]. Harun Farocki. West Germany, 1983.

Images of the World and the Inscription of War [*Bilder der Welt und Inschrift des Krieges*]. Harun Farocki. West Germany, 1989.

In Comparison. Harun Farocki. Germany, 2009.

In the Land of the War Canoes [aka *In the Land of the Head Hunters*]. Edward S. Curtis. U.S., 1914.

Jaguar. Jean Rouch. France, 1970.

La Jetée. Chris Marker. France, 1962.

The Lady with the Light Gloves [*Damen med de lyse Handsker*]. Benjamin Christensen. Denmark, 1942.

Land without Bread [*Tierra Sin Pan*]. Luis Buñuel. Spain, 1933.

The Last Laugh [*Der letzte Mann*]. F. W. Murnau. Germany, 1924.

Leaves from Satan's Book [*Blade af Satans Bog*]. Carl Theodor Dreyer. Denmark, 1921.

Leviathan. Lucien Castaing-Taylor and Véréna Paravel. U.S./U.K., 2013.

Master of the House [*Du skal ære din hustru*]. Carl Theodor Dreyer. Denmark, 1925.

Mikaël. Carl Theodor Dreyer. Germany, 1924.

Moana. Robert Flaherty. U.S., 1926.

Mockery. Benjamin Christensen. U.S., 1927.

Moi, un Noir. Jean Rouch. France, 1960.

The Mysterious Island. Benjamin Christensen / Lucien Hubbard. U.S., 1929.

The Mysterious X [aka *Sealed Orders* (*Det hemmelighedsfulde X*)]. Benjamin Christensen. Denmark, 1914.

Nanook of the North. Robert Flaherty. U.S., 1922.

Nosferatu [*Nosferatu, eine Symphonie des Grauens*]. F. W. Murnau. Germany, 1922.

The Passion of Joan of Arc [*La passion de Jeanne d'Arc*]. Carl Theodor Dreyer. France, 1928.

Petit à Petit. Jean Rouch. France, 1971.

Phantom. F. W. Murnau. Germany, 1922.

The Phantom Carriage [*Körkarlen*]. Victor Sjöström. Sweden, 1921.

Seven Footprints to Satan. Benjamin Christensen. U.S., 1929.

The Seventh Seal [*Det sjunde inseglet*]. Ingmar Bergman. Sweden, 1956.

Sweetgrass. Lucien Castaing-Taylor. U.S., 2009.

Vampyr. Carl Theodor Dreyer. Germany, 1932.

War Neuroses—Netley Hospital. Arthur Hurst. U.K., 1917.

With Captain Scott, R.N. to the South Pole. No director listed. U.K., 1912.

With Our Heroes at the Somme [*Bei unseren Helden an der Somme*]. Bild und
    Filmamt (Bufa). Germany, 1917.

Woman. Maurice Tourneur. U.S., 1918.

Airaksinen, Timo. *The Philosophy of the Marquis de Sade*. London: Routledge, 1995.

Allen, Michael, ed. and trans. *Marsilio Ficino and the Phaedran Charioteer*. Berkeley: University of California Press, 1981.

Apuleius, Lucius. *Ain schön lieblich auch kurtzweylig gedichte Lucij Apuleij von ainem gulden Esel . . . Mit schönen figuren zůgericht, grundtlich verdeutscht durch Herren Johan Sieder, etc. Ff. 71*. A. Weissenhorn: Augustae Vindelicorum, 1538.

Aquinas, St. Thomas. *Summa theologiae*. Edited by Thomas Gilby ct al. 60 vols. New York: McGraw Hill, 1964.

Asad, Talal. *Formations of the Secular: Christianity, Islam, Modernity*. Stanford, Calif.: Stanford University Press, 2003.

———. *Genealogies of Religion*. Baltimore, Md.: Johns Hopkins University Press, 1993.

Audisio, Gabriel. *The Waldensian Dissent: Persecution and Survival, c.1170–c.1570*. Cambridge, U.K.: Cambridge University Press, 1999.

Baer, Ulrich. *Spectral Evidence: The Photography of Trauma*. Cambridge, Mass.: MIT Press, 2002.

Bakhtin, Mikhail. *Rabelais and His World*. Translated by Hélène Iswolsky. Bloomington: Indiana University Press, 1984 [1968].

Barber, Malcolm. *The Cathars: Dualist Heretics in Languedoc in the High Middle Ages*. London: Longman, 2000.

Barsam, Richard. *The Vision of Robert Flaherty: The Artist as Myth and Filmmaker*. Bloomington: Indiana University Press, 1988.

Behringer, Wolfgang. *Shaman of Oberstdorf: Chonrad Stoeckhlin and the Phantoms of the Night*. Charlottesville: University Press of Virginia, 1998.

Benjamin, Walter. "Little History of Photography." In *Selected Writings, Vol. 1, Part 2: 1931–1934*, edited by Michael Jennings et al., 507–30. Cambridge, Mass.: Harvard University Press, 1999.

————. "M [The Flâneur]." In *The Arcades Project*, edited by Rolf Tiedemann, 416–55. Cambridge, Mass.: Harvard University Press, 1999.

————. "One Way Street." In *Selected Writings, Vol. 1: 1913–1926*, edited by Marcus Bullock and Michael W. Jennings, 444–88. Cambridge, Mass.: Harvard University Press, 1996.

————. "The Storyteller: Observations on the Work of Nikolai Leskov." In *Selected Writings, Vol. 3: 1935–1938*, edited by Howard Eiland and Michael W. Jennings, 143–66. Cambridge, Mass.: Harvard University Press, 2006.

————. *The Work of Art in the Age of Its Technological Reproducibility, and Other Writings on Media*. Edited by Michael W. Jennings, Brigid Doherty, and Thomas Y. Levin. Cambridge, Mass.: Harvard University Press, 2008.

Bever, Edward. "Witchcraft, Female Aggression, and Power in the Early Modern Community." *Journal of Social History* 35, no. 4 (2002): 955–88.

Binswanger, Otto. *Die Hysterie*. Wien: Hölder, 1904.

————. "Die Kriegshysterie." In *Handbuch der Ärztlichen Erfahrungen im Weltkriege, 1914–18*, edited by O. Schjerning, vol. 4: *Geistes- und Nevenkrankheiten*, edited by K. Bonhoeffer, 45–67. Leipzig: J. A. Barth, 1922.

Blümlinger, Christa. "Slowly Forming a Thought while Working on Images." In *Harun Farocki: Working on the Sightlines*, edited by Thomas Elsaesser, 163–76. Amsterdam: Amsterdam University Press, 2004.

Boddy, Janice. *Wombs and Alien Spirits: Women, Men, and the Zār Cult in Northern Sudan*. Madison: University of Wisconsin Press, 1989.

Bodin, Jean. *De la démonomanie des sorciers*. Paris: du Puys, 1580.

Bordwell, David. *The Films of Carl-Theodor Dreyer*. Berkeley: University of California Press, 1981.

————. *Making Meaning: Inference and Rhetoric in the Interpretation of Cinema*. Cambridge, Mass.: Harvard University Press, 1989.

Boguet, Henri. *An Examen of Witches*. London: Richard Clay and Sons, 1929 [1610].

Bourneville, Désiré-Magloire. *La possession de Jeanne Fery; religieuse professe du couvent des Soeurs Noires de la Ville de Mons*. Paris: Progrès Médical, 1886 [1584].

————. *Science et miracle: Louise Lateau ou la stigmatisée belge (Bibliothèque diabolique)*. Paris: Progrès Médical, 1875.

Bourneville, Désiré-Magloire, and Edmond Teinturier. *Le sabbat des sorciers (Bibliothèque diabolique)*. Paris: Progrès Médical, 1882.

Bourneville, Désiré-Magloire, Paul Régnard, Jean-Martin Charcot, and Édouard Delessert. *Iconographie photographique de la Salpêtrière: Service de M. Charcot*. 3 vols. Paris: Progrès Médical, 1877–80.

Brakhage, Stan. Lecture at the Art Institute of Chicago. John M. Flaxman Library, HFA item #10014. Audiocassette recording. Chicago, January 31, 1973.

Braun, Marta. *Picturing Time: The Work of Etienne-Jules Marey (1830–1904)*. Chicago: University of Chicago Press, 1995.

Breuer, Josef, and Sigmund Freud. *Studies on Hysteria*. London: Penguin, 1991 [1895].

Broedel, Hans Peter. *The "Malleus Maleficarum" and the Construction of Witchcraft: Theology and Popular Belief*. Manchester, U.K.: Manchester University Press, 2003.

Bruno, Giuliana. "Pleats of Matter, Folds of the Soul." In *Afterimages of Gilles Deleuze's Film Philosophy*, edited by D. N. Rodowick, 213–34. Minneapolis: University of Minnesota Press, 2010.

Bynum, Caroline Walker. *Christian Materiality: An Essay on Religion in Late Medieval Europe*. New York: Zone Books, 2011.

———. *Metamorphosis and Identity*. New York: Zone Books, 2001.

———. *Wonderful Blood: Theology and Practice in Late Medieval Northern Germany and Beyond*. Philadelphia: University of Pennsylvania Press, 2007.

Caciola, Nancy. *Discerning Spirits: Divine and Demonic Possession in the Middle Ages*. Ithaca, N.Y.: Cornell University Press, 2003.

Calder-Marshall, Arthur. *The Innocent Eye: The Life of Robert Flaherty*. New York: W. H. Allen, 1963.

Cameron, Euan. *Enchanted Europe: Superstition, Reason, and Religion, 1250–1750*. Oxford, U.K.: Oxford University Press, 2010.

Campanella, Tommaso. *The City of the Sun: A Poetical Dialogue*. Berkeley: University of California Press, 1981 [1623].

Cannon, Walter B. "'Voodoo' Death." *American Anthropologist* 44, no. 2 (1942): 169–81.

Carlino, Andrea. *Books of the Body: Anatomical Ritual and Renaissance Learning*. Chicago: University of Chicago Press, 1999.

Casebier, Allan. *Film and Phenomenology: Towards a Realist Theory of Cinematic Representation*. Cambridge, U.K.: Cambridge University Press, 1991.

Caton, Steven C. *"Lawrence of Arabia": A Film's Anthropology*. Berkeley: University of California Press, 1999.

Chanan, Michael. *The Politics of Documentary*. London: British Film Institute, 2007.

Charcot, Jean-Martin and Paul Marie Louis Pierre Richer. *Les démoniaques dans l'art: Avec 67 figures intercalées dans le texte*. Paris: Delahaye et Lecrosnier, 1887.

Christensen, Benjamin. "Benjamin Christensen's film: En studie over häxprocesserna." *Filmjournalen* 21–22 (December 25, 1921): 737.

———. "The Future of Film." In *Benjamin Christensen: An International Dane*, edited by Jytte Jensen, 38–39. New York: Museum of Modern Art, 1999.

Clark, Stuart. *Thinking with Demons: The Idea of Witchcraft in Early Modern Europe*. Oxford, U.K.: Oxford University Press, 1997.

Clifford, James. "On Ethnographic Authority." *Representations* 1, no 1 (1983): 118–46.

———. *The Predicament of Culture: Twentieth-Century Ethnography, Literature, and Art*. Cambridge, Mass.: Harvard University Press, 1988.

Cohn, Norman. *Europe's Inner Demons: An Enquiry Inspired by the Great Witch-Hunt*. London: Chatto Heinemann, 1975.

———. *The Pursuit of the Millennium: Revolutionary Millenarians and Mystical Anarchists of the Middle Ages*. London: Pimlico, 2004 [1970].

Comolli, Jean-Louis. "Filmographie commentée." In *Dreyer*, edited and translated by Mark Nash, 67. London: British Film Institute, 1977.

Conrad, Joseph. *Heart of Darkness*. New York: Norton, 1988 [1899].

Couperus, Louis. *The Hidden Force*. Amherst: University of Massachusetts Press, 1985 [1900].

Crary, Jonathan. *Techniques of the Observer: On Vision and Modernity in the Nineteenth Century*. Cambridge, Mass.: MIT Press, 1991.

Cunningham, Andrew. *The Anatomical Renaissance: The Resurrection of the Anatomical Projects of the Ancients*. Aldershot, U.K.: Scolar Press, 1997.

Dagognet, François. *Etienne-Jules Marey: A Passion for the Trace*. Translated by Robert Galeta with Jeanine Herman. New York: Zone Books, 1992.

Dalle Vacche, Angela. *Cinema and Painting: How Art Is Used in Film*. Austin: University of Texas Press, 1996.

Daly, Mary. *Gyn/ecology: The Metaethics of Radical Feminism*. Boston: Beacon, 1978.

Daston, Lorraine, and Peter Galison. *Objectivity*. New York: Zone Books, 2007.

Daston, Lorraine, and Katharine Park. *Wonders and the Order of Nature, 1150–1750*. New York: Zone Books, 1998.

De Certeau, Michel. *The Possession at Loudun*. Translated by Michael B. Smith. Chicago: University of Chicago Press, 2000 [1986].

De Lamothe-Langon, Étienne Léon. *Histoire de l'inquisition en France, depuis son établissement au XIIIe siècle, à la suite de la croisade contre les Albigeois, jusqu'en 1772, époque définitive de sa suppression*. Paris: J.-G. Dentu, 1829.

De Lancre, Pierre. *On the Inconstancy of Witches*. Tempe: Arizona Center for Medieval and Renaissance Studies, 2006 [1612].

De Moura, Manuel do Valle. *De incantantionibus seu ensalmis*. Lisbon, 1620.

De Rios, Marlene Dobkin. "The Relationship between Witchcraft Beliefs and Psychosomatic Illness." In *Anthropology and Mental Health: Setting a New Course*, edited by Joseph Westermeyer, 11–18. The Hague: Mouton, 1976.

Delaporte, François. *Anatomy of the Passions*. Translated by Susan Emanuel and edited by Todd Meyers. Stanford, Calif.: Stanford University Press, 2008.

Deleuze, Gilles. "The Brain Is the Screen." In *Two Regimes of Madness: Texts and Interviews, 1975–1995*, 282–91. New York: Semiotext(e), 2006.

———. *Cinema 1: The Movement-Image*. Translated by Hugh Tomlinson and Barbara Habberjam. Minneapolis: University of Minnesota Press, 1986.

———. *Cinema 2: The Time-Image*. Translated by Hugh Tomlinson and Robert Galeta. Minneapolis: University of Minnesota Press, 1989.

———. "Coldness and Cruelty." In *Masochism*, 9–142. New York: Zone Books, 1991 [1967].

———. *Difference and Repetition*. Translated by Paul Patton. New York: Columbia University Press, 1994 [1968].

———. *Francis Bacon: The Logic of Sensation*. Translated by Daniel W. Smith. Minneapolis: University of Minnesota Press, 2002.

Deleuze, Gilles, and Félix Guattari. *What Is Philosophy?* Translated by Hugh Tomlinson and Graham Burchell. New York: Columbia University Press, 1996 [1991].

Deren, Maya. *An Anagram of Ideas on Art, Form, and Film*. Yonkers, N.Y.: Alicat Book Shop Press, 1946. Reproduced in *Maya Deren and the American Avant-Garde*, edited by Bill Nichols. Berkeley: University of California Press, 2001.

———. *Divine Horsemen: The Living Gods of Haiti*. New York: Thames and Hudson, 1953.

Derrida, Jacques. *Archive Fever: A Freudian Impression*. Translated by Eric Prenowitz. Chicago: University of Chicago Press, 1998.

———. *Demeure: Fiction and Testimony*. Translated by Elizabeth Rottenberg. Stanford, Calif.: Stanford University Press, 2000

Didi-Huberman, Georges. *Images in Spite of All: Four Photographs from Auschwitz*. Translated by Shane B. Lillis. Chicago: University of Chicago Press, 2008 [2003].

———. *Invention of Hysteria: Charcot and the Photographic Iconography of the Salpêtrière*. Translated by Alisa Hartz. Cambridge, Mass.: MIT Press, 2003 [1982].

Dixon, Bryony. *100 Silent Films*. London: British Film Institute, 2011.

Doane, Mary Ann. *The Emergence of Cinematic Time: Modernity, Contingency, the Archive*. Cambridge, Mass.: Harvard University Press, 2002.

Drouzy, Maurice. *Carl Th. Dreyer, né Nilsson*. Paris: Editions du Cerf, 1982.

Durington, Matthew, and Jay Ruby. "Ethnographic Film." In *Made to Be Seen: Perspectives on the History of Visual Anthropology*, edited by Marcus Banks and Jay Ruby, 190–208. Chicago: University of Chicago Press, 2011.

Edelstein, Dan. *The Terror of Natural Right: Republicanism, the Cult of Nature, and the French Revolution*. Chicago: University of Chicago Press, 2010.

Eisner, Lotte. *The Haunted Screen: Expressionism in the German Cinema and the Influence of Max Reinhardt*. Berkeley: University of California Press, 1973.

Elsaesser, Thomas. *Weimar Cinema and After: Germany's Historical Imaginary.* London: Routledge, 2000.

Epstein, Julie. *Altered Conditions: Disease, Medicine, and Storytelling.* New York: Routledge, 1995.

Erasmus. *Praise of Folly.* Translated by Betty Radice. London: Penguin, 1971 [1511].

Ernst, John. *Benjamin Christensen.* Copenhagen: Danske Filmmuseum, 1967.

Evans-Pritchard, E. E. *Witchcraft, Oracles and Magic among the Azande.* Abridged ed. Oxford, U.K.: Oxford University Press, 1976 [1937].

Favret-Saada, Jeanne. *The Anti-Witch.* Translated by Matthew Carey. Chicago: HAU Books, 2015 [2009].

——. *Deadly Words: Witchcraft in the Bocage.* Cambridge, U.K.: Cambridge University Press, 1980.

Ferber, Sarah. *Demonic Possession and Exorcism in Early Modern France.* London: Routledge, 2004.

Firth, Raymond. "Introduction: Malinowski as Scientist and as Man." In *Man and Culture: An Evaluation of the Work of Bronislaw Malinowski*, edited by Raymond Firth, 1–14. London: Routledge and Kegan Paul, 1957.

Foucault, Michel. *Abnormal: Lectures at the Collège de France, 1974–1975.* Translated by Graham Burchell. New York: Picador, 2003.

——. *The Birth of the Clinic: An Archaeology of Medical Perception.* Translated by A. M. Sheridan Smith. New York: Vintage, 1973.

——. "Déviations religieuses et savoir médical." In *Hérésies et societes dans l'Europe préindustrielle, 11e–18e siècles*, edited by Jacques LeGoff, 19–25. Paris: Mouton, 1968.

——. *Madness and Civilization: A History of Insanity in the Age of Reason.* Translated by Richard Howard. New York: Vintage, 1965.

——. *The Order of Things: An Archaeology of the Human Sciences.* New York: Vintage, 1970 [1966].

Frazer, James George. *The Golden Bough: A Study in Magic and Religion.* London: Macmillan, 1976 [1880].

Freud, Sigmund. *Beyond the Pleasure Principle.* Translated by James Strachey. New York: Bantam, 1959 [1920].

——. "A Child Is Being Beaten." In *Sexuality and the Psychology of Love.* New York: Bantam, 1968 [1919].

——. *Dora: An Analysis of a Case of Hysteria.* New York: Touchstone, 1997 [1905].

——. *The Future of an Illusion.* Translated by James Strachey. New York: Norton, 1975 [1927].

——. *A General Introduction to Psychoanalysis.* New York: Horace Liveright, 1920.

————. "Mourning and Melancholy." In *The Standard Edition of the Complete Psychological Works of Sigmund Freud*, edited by James Strachey with Anna Freud, 14:243–58. London: Hogarth Press, 1961 [1917].

————. "A Note upon the 'Mystic Writing-Pad.'" In *Collected Papers*, vol. 5, edited by Philip Rieff. New York: Collier, 1963 [1925].

————. "Obsessive Actions and Religious Practices." In *The Standard Edition of the Complete Psychological Works of Sigmund Freud*, edited by James Strachey with Anna Freud, 115–27. London: Hogarth Press, 1961 [1907].

————. *On Sexuality: Three Essays on the Theory of Sexuality and Other Works*. Translated by James Strachey. London: Penguin, 1985 [1905].

————. *An Outline of Psychoanalysis*. New York: Norton, 1949 [1940].

————. *Three Case Histories: The "Wolf Man," the "Rat Man," and the Psychotic Doctor Schreber*. New York: Touchstone, 1996.

————. *Totem and Taboo: Some Points of Agreement between the Mental Lives of Savages and Neurotics*. New York: Norton, 1950 [1913].

Freytag, Gustav. *Pictures of German Life in the Fifteenth, Sixteenth, and Seventeenth Centuries*. London: Chapham and Hall, 1862.

Fried, Michael. *Why Photography Matters as Never Before*. New Haven, Conn.: Yale University Press, 2008.

Fuchs, Eduard. *Illustrierte Sittengeschichte vom Mittelalter bis zur Gegenwart*. Munich: Langen, 1912.

Fussell, Paul. *The Great War and Modern Memory*. New York: Oxford University Press, 2000.

Galassi, Peter. *Before Photography: Painting and the Invention of Photography*. New York: Museum of Modern Art, 1981.

Gaskill, Malcolm. *Witchcraft: A Very Short Introduction*. Oxford, U.K.: Oxford University Press, 2010.

Geertz, Clifford. *Works and Lives: The Anthropologist as Author*. Stanford, Calif.: Stanford University Press, 1988.

Geiler, Johann. *Die Emeis: Dis ist das buch von der Omeissen*. Strassburg: von Johannes Grieninger, 1516.

Geroulanos, Stefanos. "A Child Is Being Murdered." In *anthropologies*, edited by Richard Baxstrom and Todd Meyers, 17–30. Baltimore, Md.: Creative Capitalism, 2008.

Geschiere, Peter. *The Modernity of Witchcraft: Politics and the Occult in Postcolonial Africa*. Translated by Janet Roitman and Peter Geschiere. Charlottesville: University Press of Virginia, 1997.

————. *Witchcraft, Intimacy, and Trust: Africa in Comparison*. Chicago: University of Chicago Press, 2013.

Gilles de la Tourette, Georges. *Traité clinique et thérapeutique de l'hystérie d'après l'enseignement de la Salpêtrière*. Paris: E. Plon, Nourrit et Cie, 1891.

Ginzburg, Carlo. *The Cheese and the Worms: The Cosmos of a Sixteenth-Century Miller*. London: Routledge, 1980.

———. *Myths, Emblems, and Clues*. London: Hutchinson Radius, 1990.

———. *The Night Battles: Witchcraft and Agrarian Cults in the Sixteenth and Seventeenth Centuries*. London: Routledge, 1983.

Goldstein, Jan. *Console and Classify: The French Psychiatric Profession in the Nineteenth Century*. New York: Cambridge University Press, 1990.

———. *Hysteria Complicated by Ecstasy: The Case of Nanette Leroux*. Princeton, N.J.: Princeton University Press, 2010.

Gombrich, E. H. *The Story of Art*. New York: Phaidon, 1966.

Grafton, Anthony. *Cardano's Cosmos: The Worlds and Works of a Renaissance Astrologer*. Cambridge, Mass.: Harvard University Press, 1999.

Griaule, Marcel. *Methode de l'etnographie*. Paris: Presses Universitaires de France, 1957.

Griffiths, Alison. *Wondrous Difference: Cinema, Anthropology, and Turn-of-the-Century Visual Culture*. New York: Columbia University Press, 2002.

Grimshaw, Anna. *The Ethnographer's Eye: Ways of Seeing in Modern Anthropology*. Cambridge, U.K.: Cambridge University Press, 2001.

Guazzo, Francesco Maria. *Compendium maleficarum*. Translated by E. A. Ashwin and edited by Montague Summers. New York: Dover, 1988 [1929].

Gunning, Tom. "An Aesthetic of Astonishment: Early Film and the (In)Credulous Spectator." In *Viewing Positions: Ways of Seeing Film*, edited by Linda Williams, 114–33. New Brunswick, N.J.: Rutgers University Press, 1995.

Haggard, H. Rider. *King Solomon's Mines*. London: Cassell and Company, 1885.

Hammond, Paul, ed. *The Shadow and Its Shadow: Surrealist Writings on the Cinema*. 3rd ed. San Francisco: City Lights Books, 2000.

Hansen, Joseph. *Zauberwahn, Inquisition und Hexenprozess im Mittelalter und die Entstehung der grossen Hexenverfolgung*. Munich: R. Oldenbourg, 1900.

Harbord, Alice. *Chris Marker: "La Jetée."* London: Afterall Books, 2009.

Hardy, Forsyth. *Grierson on Documentary*. London: Collins, 1946.

Heinemann, Franz. *Der Richter und die Rechtspflege in der deutschen Vergangeheit: Mit 159 Abbildungen und Beilagen nach den Originalen aus dem fünfzehnten bis achtzehnten Jahrhundert*. Leipzig: Eugen Diederichs, 1900.

Hippocratic Writings. Translated by J. Chadwick and W. N. Mann. London: Penguin, 1983.

Hollier, Denis. "The Question of Lay Ethnography [The Entropological Wild Card]." In *Undercover Surrealism: Georges Bataille and* DOCUMENTS, edited by Dawn Ades and Simon Baker, 58–64. Cambridge, Mass.: MIT Press, 2006.

Huneman, Philippe. "Writing the Case: Pinel as Psychiatrist." *Republics of Letters: A Journal for the Study of Knowledge, Politics, and the Arts*, 3, no. 2 (2014): 1–28.

Jackson, Kevin. *Constellation of Genius: 1922, Modernism Year One.* New York: Farrar, Straus and Giroux, 2013.

Jacob, Margaret C. "The Materialist World of Pornography." In *The Invention of Pornography: Obscenity and the Origins of Modernity, 1500–1800*, edited by Lynn Hunt, 157–202. New York: Zone Books, 1993.

James, King of England. *Daemonologie.* In *The Demonology of King James I: Includes the Original Text of Daemonologie and News from Scotland.* Woodbury, Minn.: Llewellyn, 2011 [1597].

James, William. *Proceedings of the American Society for Psychical Research* 23 (1909).

Janet, Pierre. *Névroses et idées fixes.* Paris: Félix Alcan, 1898.

Jeanne des Anges, Sister. *Soeur Jeanne des Anges, supérieure des Ursulines de Loudun, XVIIe siècle: Autobiographie d'une hystérique possédée, d'après le manuscrit inédit de la bibliothèque de Tours (Bibliothèque diabolique).* Paris: Progrès Médical, 1886.

Jensen, Jytte. "Benjamin Christensen: An International Dane." In *Benjamin Christensen: An International Dane*, edited by Jytte Jensen, 4–6. New York: Museum of Modern Art, 1999.

Kaes, Anton. *Shell Shock Cinema: Weimar Culture and the Wounds of War.* Princeton, N.J.: Princeton University Press, 2009.

Kaufmann, Doris. "Science as Cultural Practice: Psychiatry in the First World War and Weimar Germany." *Journal of Contemporary History* 34, no. 1 (1999): 125–44.

Keller, Corey. "Sight Unseen: Picturing the Invisible." In *Brought to Light: Photography and the Invisible, 1840–1900*, edited by Corey Keller, 19–35. San Francisco: San Francisco Museum of Modern Art, 2008.

Kieckhefer, Richard. *European Witch Trials: Their Foundations in Popular and Learned Culture, 1300–1500.* London: Routledge, 1976.

———. *Magic in the Middle Ages.* Cambridge, U.K.: Cambridge University Press, 1989.

Kiel, Charlie. "Steel Engines and Cardboard Rockets: The Status of Fiction and Nonfiction in Early Cinema." In *F Is for Phony: Fake Documentary and Truth's Undoing*, edited by Alexandra Juhasz and Jesse Lerner, 39–49. Minneapolis: University of Minnesota Press, 2006.

Klein, Melanie. "A Contribution to the Psychogenesis of Manic-Depressive States" [1934]. In *Contributions to Psycho-Analysis, 1921–1945.* London: Hogarth Press, 1950.

Klibansky, Raymond, Erwin Panofsky, and Fritz Saxl. *Saturn and Melancholy: Studies on the History of Natural Philosophy, Religion, and Art.* New York: Basic Books, 1964.

Koerner, Joseph Leo. "Hieronymus Bosch's World Picture." In *Picturing Science, Producing Art*, edited by Peter Galison and Carolyn Jones, 297–323. New York: Routledge, 1998.

————. *The Moment of Self-Portraiture in German Renaissance Art.* Chicago: University of Chicago Press, 1997.

————. *The Reformation of the Image.* Chicago: University of Chicago Press, 2004.

Kors, Alan C., and Edward Peters. "Scepticism, Doubt, and Disbelief in the Sixteenth and Seventeenth Centuries" (introduction to Part 7). In *Witchcraft in Europe, 1100–1700: A Documentary History*, edited by Alan C. Kors and Edward Peters, 311–13. Philadelphia: University of Pennsylvania Press, 1972.

Koyré, Alexandre. *From the Closed World to the Infinite Universe.* Baltimore, Md.: Johns Hopkins University Press, 1957.

Krafft-Ebing, Richard von. *Psychopathia Sexualis: Eine Klinisch-Forensische Studie.* Stuttgart: Enke, 1886.

Lamothe-Langon, Étienne-Léon de. *Histoire de l'inquisition en France, depuis son établissement au XIIIe siècle, à la suite de la croisade contre les Albigeois, jusqu'en 1772, époque définitive de sa suppression.* Paris: J.-G. Dentu, 1829.

Laplanche, J., and J.-B. Pontalis. *The Language of Psycho-Analysis.* London: Hogarth Press, 1973.

Largier, Niklaus. *In Praise of the Whip: A Cultural History of Arousal.* Translated by Graham Harman. New York: Zone Books, 2007.

Larner, Christina. *Witches and Religion: The Politics of Popular Belief.* Oxford, U.K.: Blackwell, 1984.

Latour, Bruno. *On the Modern Cult of the Factish Gods.* Durham, N.C.: Duke University Press, 2010.

Lecky, W.E.H. *History of the Rise and Influence of the Spirit of Rationalism in Europe.* New York: Georges Braziller, 1955 [1865].

Lefebvre, Henri. *The Production of Space.* Translated by Donald Nicholson-Smith. Oxford, U.K.: Blackwell, 1991.

Lehmann, Alfred. *Hypnosen og de dermed beslaegtede normale Tilstande.* Copenhagen: Philipsen, 1890.

————. *Overtro og trolddom fra de aeldste tider til vore dage.* Copenhagen: J. Frimodts Forlag, 1893–96.

Lerner, Paul. *Hysterical Men: War, Psychiatry, and the Politics of Trauma in Germany, 1890–1930.* Ithaca, N.Y.: Cornell University Press, 2009.

Levack, Brian P. *The Witch-Hunt in Early Modern Europe.* 3rd ed. London: Longman, 2006.

Levi, Anthony. *Renaissance and Reformation: The Intellectual Genesis.* New Haven, Conn.: Yale University Press, 2002.

Lévi-Strauss, Claude. *Structural Anthropology.* Translated by John and Doreen Weightman. New York: Basic Books, 1992 [1955].

————. *Tristes tropiques.* Translated by John Weightman and Doreen Weightman. New York: Penguin, 1992 [1955].

Lévy-Bruhl, Lucien. *How Natives Think.* London: Allen and Unwin, 1910.

Low, Rachael. *History of the British Film, Vol. 2: 1906–1914.* London: Routledge, 1997 [1949].

Lunde, Arne. "Benjamin Christensen in Hollywood." In *Benjamin Christensen: An International Dane,* edited by Jytte Jensen, 23–33. New York: Museum of Modern Art, 1999.

———. "The Danish Sound Feature at Nordisk." In *Benjamin Christensen: An International Dane,* edited by Jytte Jensen, 34–37. New York: Museum of Modern Art, 1999.

———. "Scandinavian Auteur as Chameleon: How Benjamin Christensen Reinvented Himself in Hollywood." *Journal of Scandinavian Cinema* 1, no. 1 (2010): 7–23.

Luhrmann, T. M. "A Hyperreal God and Modern Belief: Toward an Anthropological Theory of Mind." *Current Anthropology* 53, no. 4 (2012): 371–95.

———. *Persuasions of the Witch's Craft: Ritual Magic in Contemporary England.* Cambridge, Mass.: Harvard University Press, 1989.

MacDougall, David. *The Corporeal Image: Film, Ethnography, and the Senses.* Princeton, N.J.: Princeton University Press, 2006.

Macfarlane, Alan. *Witchcraft in Tudor and Stuart England: A Regional and Comparative Study.* New York: Harper and Row, 1970.

MacIntyre, Alasdair. *After Virtue.* Notre Dame, Ind.: University of Notre Dame Press, 1981.

Mackay, Christopher S. *The Hammer of Witches: A Complete Translation of the "Malleus Maleficarum."* Cambridge, U.K.: Cambridge University Press, 2009 [1487].

Maggi, Armando. *Satan's Rhetoric: A Study of Renaissance Demonology.* Chicago: University of Chicago Press, 2001.

Malabou, Catherine. *The New Wounded: From Neurosis to Brain Damage.* Translated by Steven Miller. New York: Fordham University Press, 2012.

Malinowski, Bronislaw. *Argonauts of the Western Pacific: An Account of Native Enterprise and Adventure in the Archipelagoes of Melanesian New Guinea.* London: Routledge, 2002 [1922].

———. *A Diary in the Strict Sense of the Term.* London: Routledge and Kegan Paul, 1967.

Marrati, Paola. *Gilles Deleuze: Cinema and Philosophy.* Translated by Alisa Hartz. Baltimore, Md.: Johns Hopkins University Press, 2008.

Maspero, Gaston. *Egyptian Archaeology.* New York: G. P. Putnam's Sons, 1887.

Métraux, Alfred. *Voodoo in Haiti.* New York: Schocken, 1972 [1958].

Meyer, Birgit, and Dick Houtman. "Introduction: Material Religion—How Things Matter." In *Things: Religion and the Question of Materiality,* edited by Dick Houtman and Birgit Meyers, 1–25. New York: Fordham University Press, 2012.

Micale, Mark S. "Hysteria and Its Historiography: A Review of Past and Present Writings (I and II)." *History of Science* 27, no. 77 (1989): 224–61, 319–51.

———. *Hysterical Men: The Hidden History of Male Nervous Illness*. Cambridge, Mass.: Harvard University Press, 2008.

Michaud, Philippe-Alain. *Aby Warburg and the Image in Motion*. Translated by Sophie Hawkes and introduction by Georges Didi-Huberman. New York: Zone Books, 2004.

Mitchell, Stephen A. *Witchcraft and Magic in the Nordic Middle Ages*. Philadelphia: University of Pennsylvania Press, 2010.

Mitchell, W.J.T. *What Do Pictures Want? The Lives and Loves of Images*. Chicago: University of Chicago Press, 2005.

Monter, E. William. "Scandinavian Witchcraft in Anglo-American Perspective." In *Early Modern European Witchcraft: Centres and Peripheries*, edited by Bengt Ankarloo and Gustav Henningsen, 425–34. Oxford, U.K.: Oxford University Press, 1993.

Monty, Ib. "Benjamin Christensen in Germany: The Critical Reception of His Films in the 1910s and 1920s." In *Nordic Explorations: Film before 1930*, edited by John Fullerton and Jan Olsson, 41–55. Sydney: John Libbey, 1999.

Müller, Friedrich Max. *Lectures on the Science of Religion*. New York: Charles Scribner and Company, 1872.

Murphy, Timothy S. *Wising Up the Marks: The Amodern William Burroughs*. Berkeley: University of California Press, 1997.

Murray, Margaret Alice. *The Witch-Cult in Western Europe: A Study in Anthropology*. Oxford, U.K.: Clarendon Press, 1921.

Nash, Mark. *Dreyer*. London: British Film Institute, 1977.

Nichols, Bill. *Representing Reality: Issues and Concepts in Documentary*. Bloomington: Indiana University Press, 1991.

Nider, Johannes. *Formicarius*. Bk. 5 repr. in *Malleus Maleficarum* (1669 ed.), ca. 1435–37.

Nietzsche, Friedrich. *The Gay Science*. Translated by Walter Kaufmann. Cambridge, U.K.: Cambridge University Press, 2001 [1882].

Oesterreich, Traugott K. *Occultism and Modern Science*. London, Methuen, 1923.

Owen, A. R. G. *Hysteria, Hypnosis and Healing: The Work of J.-M. Charcot*. London, Dobson, 1971.

Ozment, Steven. *The Serpent and the Lamb: Cranach, Luther, and the Making of the Reformation*. New Haven, Conn.: Yale University Press, 2011.

Park, Katharine. *Secrets of Women: Gender, Generation, and the Origins of Human Dissection*. New York: Zone Books, 2006.

Perkins, Judith. *The Suffering Self: Pain and Narrative in the Early Christian Era*. New York: Routledge, 1995.

Petherbridge, Deanna. *Witches and Wicked Bodies*. Edinburgh: National Galleries of Scotland, 2013.

Phillips, Henry Albert. *The Photodrama*. New York: Arno Press, 1914.

Pinney, Christopher. *Photography and Anthropology*. London: Reaktion, 2011.

Plas, Régine. *Naissance d'une science humaine: La psychologie*. Paris: Presses Universitaires de Rennes, 2000.

Porter, Roy. "Review Article: Seeing the Past." *Past and Present* 118 (Feb. 1988): 186–205.

Prierias, Sylvester. *De strigimagarum daemonumque mirandis*. Rome, 1575.

Prodger, Phillip. *Time Stands Still: Muybridge and the Instantaneous Photography Movement*. Oxford, U.K.: Oxford University Press, 2003.

Purkiss, Diane. *The Witch in History: Early Modern and Twentieth-Century Representations*. London: Routledge, 1996.

Rabelais, François. *Gargantua and Pantagruel*. Translated by M. A. Screech. London: Penguin, 2006 [1532–64].

Rajchman, John. *The Deleuze Connections*. Cambridge, Mass.: MIT Press, 2000.

Rawlinson, George. *History of Ancient Egypt*. London: Longman and Green, 1881.

Reverdy, Pierre. "L'image." In *Nord-Sud, Self-Defence et autres écrits sur l'art et poésie, 1917–1926*, 73–75. Paris: Flammarion, 1975.

Reynolds, Pamela. *Traditional Healers and Childhood in Zimbabwe*. Athens: Ohio University Press, 1996.

Richter, Gerhard. *Atlas*. New York: Distributed Art Publishers, 2006.

Rivers, W.H.R. *Conflict and Dream*. London: Kegan Paul, 1923.

———. *The History of Melanesian Society: Percy Sladen Trust Expedition to Melanesia*. 2 vols. Cambridge, U.K.: Cambridge University Press, 1914.

———. *Instinct and the Unconscious*. Cambridge, U.K.: Cambridge University Press, 1920.

———. *Kinship and Social Organization*. London: Constable, 1914.

———. *Medicine, Magic and Religion*. London: Routledge, 2001 [1924].

———. "The Repression of War Experience." *Proceedings of the Royal Society of Medicine* (Psychiatry Section) 11 (1918): 1–20.

———. *The Todas*. London: Macmillan, 1906.

Robisheaux, Thomas Willard. *The Last Witch of Langenburg: Murder in a German Village*. New York: Norton, 2009.

Rogers, Katharine M. *The Troublesome Helpmate: A History of Misogyny in Literature*. Seattle: University of Washington Press, 1966.

Roper, Lyndal. *Oedipus and the Devil: Witchcraft, Sexuality, and Religion in Early Modern Europe*. London: Routledge, 1994.

————. *Witch Craze: Terror and Fantasy in Baroque Germany.* New Haven, Conn.: Yale University Press, 2004.

Ronell, Avital. *The Telephone Book: Technology, Schizophrenia, Electric Speech.* Lincoln: University of Nebraska Press, 1991.

————. *The Test Drive.* Urbana: University of Illinois Press, 2005.

Rotha, Paul. *Robert J. Flaherty: A Biography.* Edited by Jay Ruby. Philadelphia: University of Pennsylvania Press, 1983.

Rowland, Robert. "'Fantasticall and Devilishe Persons': European Witch-Beliefs in Comparative Perspective." In *Early Modern European Witchcraft: Centres and Peripheries,* edited by Bengt Ankarloo and Gustav Henningsen, 161–90. Oxford, U.K.: Oxford University Press, 1993.

Rudkin, David. *"Vampyr."* London: British Film Institute, 2005.

Russell, Catherine. "Surrealist Ethnography: *Las Hurdes* and the Documentary Unconscious." In *F Is for Phony: Fake Documentary and Truth's Undoing,* edited by Alexandra Juhasz and Jesse Lerner, 99–115. Minneapolis: University of Minnesota Press, 2006.

Russell, Jeffrey Burton. *Lucifer: The Devil in the Middle Ages.* Ithaca, N.Y.: Cornell University Press, 1986.

————. *Witchcraft in the Middle Ages.* Ithaca, N.Y.: Cornell University Press, 1972.

Sacher-Masoch, Leopold von. *Venus in Furs.* In *Masochism,* 143–271. New York: Zone Books, 1991 [1870].

Sade, Marquis de. *The 120 Days of Sodom, and Other Writings.* Translated by Austryn Wainhouse and Richard Seaver. New York: Grove Press, 1966.

Schedel, Hartmann. *Liber Chronicarum.* Augsburg, 1497.

Schmidgen, Henning. *Bruno Latour in Pieces: An Intellectual Biography.* Translated by Gloria Custance. New York: Fordham University Press, 2015 [2011].

Schmitt, Jean-Claude. *Le corps, les rites, les rêves, le temps: Essais d'anthropologie medieval.* Paris: Gallimard, 2001.

Scholem, Gershom. *Major Trends in Jewish Mysticism.* New York: Schocken, 1995 [1941].

Shakespeare, William. *Macbeth.* London: Penguin, 2000 [1606].

Sharpe, Jim. "Women, Witchcraft, and the Legal Process." In *Women, Crime, and the Courts in Early Modern England,* edited by Jennifer Kermode and Garthine Walker, 113–30. Chapel Hill: University of North Carolina Press, 1994.

Sherry, Norman. *Conrad's Western World.* Cambridge, U.K.: Cambridge University Press, 1980.

Siegel, James. *Naming the Witch.* Stanford, Calif.: Stanford University Press, 2006.

Slobodin, Richard. *W.H.R. Rivers: Pioneer, Anthropologist, Psychiatrist of "The Ghost Road."* Gloucestershire, U.K.: Sutton, 1997 [1978].

Smith, Moira. "The Flying Phallus and the Laughing Inquisitor: Penis Theft in the Malleus Maleficarum." *Journal of Folklore Research* 39, no. 1 (2002): 85–117.

Snell, Otto. *Hexenprozesse und Geistesstörung.* Munich: J. F. Lehmann, 1891.

Sollier, Paul Auguste. *Genèse et nature de l'hystérie, recherches cliniques et expérimentales de psycho-physiologie.* Paris: Félix Alcan, 1897.

Solnit, Rebecca. *River of Shadows: Eadweard Muybridge and the Technological Wild West.* New York: Viking, 2003.

Sperber, Dan. *On Anthropological Knowledge.* Cambridge, U.K.: Cambridge University Press, 1982.

Spina, Alphonsus de. *Fortalitium fidei.* Strasbourg, ca. 1471.

Starhawk. *The Spiral Dance: A Rebirth of the Ancient Religion of the Great Goddess.* San Francisco: Harper and Row, 1979.

Stephens, Walter. *Demon Lovers: Witchcraft, Sex, and the Crisis of Belief.* Chicago: University of Chicago Press, 2002.

Stevenson, Jack. *Witchcraft through the Ages: The Story of "Häxan," the World's Strangest Film and the Man Who Made It.* Goldaming, U.K.: FAB Press, 2006.

Stocking, George W. *After Tylor: British Social Anthropology, 1888–1951.* Madison: University of Wisconsin Press, 1995.

———. "The Ethnographer's Magic: Fieldwork in British Social Anthropology from Tylor to Malinowski." In *Observers Observed: Essays on Ethnographic Fieldwork*, edited by George W. Stocking, vol. 1 of *History of Anthropology*, 70–120. Madison: University of Wisconsin Press, 1983.

———. *The Ethnographer's Magic, and Other Essays in the History of Anthropology.* Madison: University of Wisconsin Press, 1992.

Strauss, Jonathan. *Human Remains: Medicine, Death, and Desire in Nineteenth-Century Paris.* New York: Fordham University Press, 2012.

Strauven, Wanda, ed. *The Cinema of Attractions, Reloaded.* Amsterdam: Amsterdam University Press, 2006.

Suhr, Christian, and Rane Willerslev. "Can Film Show the Invisible? The Work of Montage in Ethnographic Filmmaking." *Current Anthropology* 53, no. 3 (2012): 282–301.

Summers, Montague, ed. *The Discovery of Witches: A Study of Master Matthew Hopkins Commonly Called Witch Finder General 1647.* Whitefish, Mont.: Kessinger, 2010.

Sylvester, David. *Interviews with Francis Bacon.* New York: Thames and Hudson, 1987.

Szabari, Antónia. *Less Rightly Said: Scandals and Readers in Sixteenth-Century France.* Stanford, Calif.: Stanford University Press, 2009.

————. "The Scandal of Religion: Luther and Public Speech in the Reformation." In *Political Theologies: Public Religion in a Post-secular World*, edited by Hent de Vries and Lawrence E. Sullivan, 122–36. New York: Fordham University Press, 2006.

Thomas, Keith. *Religion and the Decline of Magic*. New York: Scribner, 1971.

Tosh, Nick. "Possession, Exorcism and Psychoanalysis." *Studies in History and Philosophy of Biological and Biomedical Sciences* 33 (2002): 583–96.

Tucker, Jennifer. *Nature Exposed: Photography as Eyewitness in Victorian Science*. Baltimore, Md.: Johns Hopkins University Press, 2005.

Tybjerg, Casper. Audio commentary. *Häxan: Witchcraft through the Ages*. New York: Criterion Collection, 2001.

————. "Images of the Master." In *Benjamin Christensen: An International Dane*, edited by Jytte Jensen, 8–22. New York: Museum of Modern Art, 1999.

————. "Red Satan: Carl Theodor Dreyer and the Bolshevik Threat." In *Nordic Explorations: Film before 1930*, edited by John Fullerton and Jan Olsson, 19–40. Sydney: John Libbey, 1999.

Tylor, Edward Burnett. *Primitive Culture: Researches into the Development of Mythology, Philosophy, Religion, Language, Art, and Custom*. 2 vols. London: John Murray, 1871.

Vaneigem, Raoul. *The Movement of the Free Spirit*. New York: Zone Books, 1994 [1986].

Wahlberg, Malin. *Documentary Time: Film and Phenomenology*. Minneapolis: University of Minnesota Press, 2008.

Walker, Alison Tara. "The Sound of Silents: Aurality and Medievalism in Benjamin Christensen's *Häxan*." In *Mass Market Medieval: The Middle Ages in Popular Culture*, edited by David W. Marshall, 42–56. Jefferson, N.C.: McFarland, 2007.

Walker, D. P. *Spiritual and Demonic Magic from Ficino to Campanella*. London: Warburg Institute, 1958.

Warburg, Aby M. *Images from the Region of the Pueblo Indians of North America*. Ithaca, N.Y.: Cornell University Press, 1995.

Weber, Max. *The Protestant Ethic and the Spirit of Capitalism*. Translated by Talcott Parsons. London: Routledge, 2001 [1930].

Weyer, Johann. *De praestigiis daemonium*. Edited by George Mora. Binghamton, N.Y.: Medieval and Renaissance Texts and Studies, 1991 [1563].

Winston, Brian. *Claiming the Real II: Documentary: Grierson and Beyond*. London: British Film Institute, 2008.

Yates, Frances. *The Art of Memory*. London: Pimlico, 1992 [1966].

————. *Giordano Bruno and the Hermetic Tradition*. London: Routledge, 2002 [1964].

————. *The Occult Philosophy in the Elizabethan Age*. London: Routledge, 2001 [1979].

Young, Michael. *Malinowski's Kiriwina: Fieldwork Photography, 1915–1918*. Chicago: University of Chicago Press, 1998.

Zika, Charles. *The Appearance of Witchcraft: Print and Visual Culture in Sixteenth-Century Europe*. London: Routledge, 2008.

————. *Exorcising Our Demons: Magic, Witchcraft and Visual Culture in Early Modern Europe*. Leiden: Brill, 2003.

forms of living

Stefanos Geroulanos and Todd Meyers, *series editors*

Georges Canguilhem, *Knowledge of Life*. Translated by Stefanos Geroulanos and Daniela Ginsburg, Introduction by Paola Marrati and Todd Meyers.

Henri Atlan, *Selected Writings: On Self-Organization, Philosophy, Bioethics, and Judaism*. Edited and with an Introduction by Stefanos Geroulanos and Todd Meyers.

Catherine Malabou, *The New Wounded: From Neurosis to Brain Damage*. Translated by Steven Miller.

François Delaporte, *Chagas Disease: History of a Continent's Scourge*. Translated by Arthur Goldhammer, Foreword by Todd Meyers.

Jonathan Strauss, *Human Remains: Medicine, Death, and Desire in Nineteenth-Century Paris*.

Georges Canguilhem, *Writings on Medicine*. Translated and with an Introduction by Stefanos Geroulanos and Todd Meyers.

François Delaporte, *Figures of Medicine: Blood, Face Transplants, Parasites*. Translated by Nils F. Schott, Foreword by Christopher Lawrence.

Juan Manuel Garrido, *On Time, Being, and Hunger: Challenging the Traditional Way of Thinking Life*.

Pamela Reynolds, *War in Worcester: Youth and the Apartheid State*.

Vanessa Lemm and Miguel Vatter, eds., *The Government of Life: Foucault, Biopolitics, and Neoliberalism*.

Henning Schmidgen, *The Helmholtz Curves: Tracing Lost Time*. Translated by Nils F. Schott.

Henning Schmidgen, *Bruno Latour in Pieces: An Intellectual Biography*. Translated by Gloria Custance.

Veena Das, *Affliction: Health, Disease, Poverty.*

Kathleen Frederickson, *The Ploy of Instinct: Victorian Sciences of Nature and Sexuality in Liberal Governance.*

Roma Chatterji, ed., *Wording the World: Veena Das and Scenes of Inheritance.*

Jean-Luc Nancy and Aurélien Barrau, *What's These Worlds Coming To?* Translated by Travis Holloway and Flor Méchain. Foreword by David Pettigrew.

Anthony Stavrianakis, Gaymon Bennett, and Lyle Fearnley, eds., *Science, Reason, Modernity: Readings for an Anthropology of the Contemporary.*

Richard Baxstrom and Todd Meyers, *Realizing the Witch: Science, Cinema, and the Mastery of the Invisible.*

Hervé Guibert, *Cytomegalovirus: A Hospitalization Diary.* Introduction by David Caron, Afterword by Todd Meyers, Translated by Clara Orban.